THE ECOLOGY OF THE ALPINE ZONE OF MOUNT KENYA

MONOGRAPHIAE BIOLOGICAE

EDITOR
P. VAN OYE
Gent

VOLUMEN XVII

DR. W. JUNK PUBLISHERS - THE HAGUE - 1967

THE ECOLOGY OF THE ALPINE ZONE OF MOUNT KENYA

by

MALCOLM JAMES COE

DR. W. JUNK PUBLISHERS - THE HAGUE - 1967

Printed in the Netherlands
Zuid-Nederlandsche Drukkerij N.V. — 's-Hertogenbosch

The main peaks of Mount Kenya. The highest points Batian and Nelion are in the centre holding the Diamond glacier between then. The Darwin glacier with its recent moraines are directly below the main peaks and the Heim and Forel Glaciers on the left.

LIST OF CONTENTS

The monograph is based, primarily, on field work carried out on Mount Kenya between December, 1957 and January, 1963, and was used in part as a Ph. D. thesis of the University of London.

A brief account of the history of the mountain's exploration is followed by an outline of the physiography, geology and glacial geology, particularly where they are relevant to the study of the ecology of the Alpine Zone.

Some confusion has arisen over the delimitation of Alpine vegetation zones and these are discussed in the light of the Author's observations. Following a brief description of the Ericaceous Zone, the main plant communities of the Alpine Zone are described and are found to be more closely related to the mountain's physiography than to factors of altitude.

The climate in conjunction with the mountain's glacially eroded surface are shown to be the primary soil-forming and distributing factors, while the marked diurnal temperature fluctuations are largely responsible for the great predominance of rosette growth forms amongst Alpine plant communities. Frost soil phenomena are also shown to be of great importance in the establishment of plant communities.

The presence of a distinct Nival Zone close to the glaciers presents an ideal site for the study of the primary stages of plant colonisation and soil formation. It is shown that the vegetation in this region supports glacio-geological opinion that Mount Kenya's glaciers have undergone a period of advance in recent time. Other stages of vegetation development at lower altitudes are also described.

The vertebrate fauna is shown to exhibit a remarkable spectrum of food preference and habitat which, together with an apparent reduction of the number of young produced, is an important factor in preserving the essentially slow-growing Alpine vegetation.

The features of major importance are discussed and suggestions are made for a further programme of work in the Alpine Zone of Mount Kenya.

ACKNOWLEDGEMENTS

The work described here was started when the Author was included in the International Geophysical Year Mount Kenya Expedition under the leadership of Professor I.S. LOUPEKINE. I am grateful to the leader of this expedition and to other members who stimulated my interests in this work, and offered their companionship on numerous long and often uncomfortable safaris on the mountain.

I would like to express my thanks to Dr. Olov HEDBERG, of the Department of Systematic Botany, Uppsala, Sweden, for his advice and help in identifying my earlier plant collections; also to Professor Robert FRIES of Stockholm, for supplying me with reprints of his valuable papers on this and other East African mountains.

In the preparation of this manuscript I wish to express my gratitude to the late Professor W. H. PEARSALL, Dr. George SALT, and Professor Carl TROLL for their help and encouragement.

In Nairobi I have been helped in many ways by Dr. B. VERDCOURT and Miss D. NAPPER of the East African Herbarium, by Bob CARCASSON and John WILLIAMS of the Coryndon Museum, and by my colleagues, John SALE in the Department of Zoology and Dr. W. T. W. MORGAN, Head of the Department of Geography in the University College, Nairobi with many useful discussions and criticisms. I am also indebted to the following members of Government Departments in Kenya who have helped by supplying unpublished information: Dick JONES of the Kenya Hydrology Department for information concerning the rainfall of Mount Kenya; the Director, Kenya Meteorological Department for valuable data on temperatures; Frank CHARNLEY of the Kenya Survey Department for help with maps; the Director of Hunting Clan Aero Surveys for permission to use their aerial photographs of the mountain; the Directorate of Overseas Surveys for supplying lay-downs of the 1948 R.A.F. aerial photographs; the Soil Chemists of the Scott Agricultural Laboratories for help with Alpine soils; and my very good friend, Brian BAKER, for his invaluable help with geological and related matters.

I wish to express my thanks to the University College for grants made to assist with porter and other expenses when visiting the mountain.

Among many other colleagues who have assisted me in various ways I should like to mention Dr. G. S. NELSON, of the London School of Hygiene and Tropical Medicine, and Dr. Heinz LOEFFLER, of Vienna University, whose council and advice have been of great use and who shared my enthusiasm for this wonderful mountain.

I am grateful to Pierre LOUIS and Mr. DHANJAL, of the University

College, for their assistance in the preparation of plates, also to Mr. D'Souza for his careful draftsmanship in improving some of my maps. Last, but by no means least, I would like to thank Mrs. Dearden for her care in typing the manuscript, and my wife, Unity, for her encouragement in preparing the final draft for this manuscript.

All field work conducted on Mount Kenya has been carried out with the kind permission of the Director of the Kenya National Parks, Mr Mervyn Cowie, and the ever willing cooperation of his Mountain Park Wardens, John Alexander and Bill Woodley.

INTRODUCTION

For centuries the peak of Mount Kenya has held a magical and religious significance for the Bantu and Nilohamitic peoples around its base. The Kikuyu live around the Eastern and Southern boundaries and the closely related Uembu and Umeru on the S.E. and N.E. respectively. Early in this century the Masai lived to the N.W. and North, but after continual warfare between them and their neighbours, the European administrators of that time moved them to a special reserve to the South, which accounts at the present day for the retention in the Masai language of many words that refer to Mount Kenya.

Kikuyu folk-lore tells how, when the earth was formed, a man named Mogai made a great mountain, *Kere-Nyaga*. The fine white powder (snow) covering the peak, which they called *ira*, was said to be the bed of Ngai (God), and during male and female circumcision ceremonies a white powder was placed on the wound, and the initiates were told that this material had been brought from the summit of the mountain. In fact all important tribal ceremonies were, and in many cases still are conducted facing the mountain. Such occasions include marriage and sacrifice when, in time of hardship, Ngai's aid is called upon (CAGNOLO 1933, KENYATTA 1938, CHIRA 1959).

Other tribes who could see the mountain gave it names which in many cases were very similar to the Kikuyu. The Wakamba called it Kima ya Kegnia, while other tribes knew it as Ndur (Kimaja) Kegnia and Kirenia. The Masai and Wanderobo names were simpler and purely descriptive, being Doinyo Ebor (white mountain) and Doinyo Egeri (black mountain) respectively. The difference in the two latter names (i.e. white and black) is more easily understood when the mountain has been viewed from all sides for some aspects, being devoid of glaciers, usually appear black.

Discovery and Exploration.

1849. The first European to see Mount Kenya was a German missionary, John Ludwig KRAPF, who with his companion REBMANN had heard many stories of snow-capped mountains in the interior. There is little doubt that Arab traders had used Kilimanjaro and Mount Kenya as land marks for their slaving and trading caravans for decades before. REBMANN discovered Kilimanjaro in 1848, and on December 3rd, 1849, KRAPF saw what he described as "two large horns or pillars, as it were rising over an enormous mountain to the northwest of Kilimanjaro, covered with a white substance" from a hill near Kitui (KRAPF 1860, DUTTON 1929,

RICHARDS & PLACE 1960). He was fortunate enough on this journey to meet an Uembu, Rumu wa Kikandi, who lived at the foot of the mountain. He told him that the kirira (snow) that covered the summit rolled down the mountain with a great noise, and that the water flowed into a great lake that fed the Dana (Tana) River. KRAPF concluded that the peaks were probably glaciated. It is difficult to decide to which lake Kikandi was referring, as there are no lakes fed from mountain streams below the montane forests upper limit. Since there is evidence (MACKINDER 1900) that Wanderobo lived, or at least visited the Alpine Zone, it seems possible that the Uembu at this time were also familiar with the Alpine lakes (probably lake Ellis or Michaelson on this side of the mountain).

In spite of KRAPF's vivid account of his observations COOLEY (1852), a distinguished geographer of the period, dismissed both his and REBMANN's experiences as halucinations.

1877. HILDEBRANT heard stories describing the mountain while he was on a plant collecting expedition in the Kitui area, but he did not actually see it.

1883. Joseph THOMSON (1885) was the next European to see Mount Kenya, and he approached it from the westerly aspect. He marched through Masailand from Kilimanjaro and after passing Donyo Longonot he proceeded northwards over Kinangop, where he sighted a range of mountains, Subugu le Poron, which he named Aberdare after the President of the Royal Geographical Society. From here the expedition walked via Thomson's Falls to the western edge of the Mount Kenya forest. Unfortunately trouble with local tribes forced them to abandon any attempt to climb the mountain, so they turned North and made their way to Lake Baringo, in the West.

1885. Shortly after THOMSON's visit to the mountain, Count Samuel TELEKI succeeded in traversing the dense montane forest, and thus he became the first white man to reach the Alpine Zone. In his account he notes the Giant Lobelias and Senecios and other important features. From VON HÖHNEL's account of the expedition it seems that they trekked up the ridge of the deep valley that GREGORY later named in honour of Count Teleki, to an altitude of approximately 15,000 feet. Since the top of this valley is a bowl-shaped cwm, he formed the mistaken impression that this region was the crater of a large volcano (VON HÖHNEL 1894). The explorers returned to the eastern side of the mountain in 1892, but they did not on this occasion get through the forest.

1893. Nearly ten years later the great explorer, geologist and naturalist, Dr. J. W. GREGORY, who was the first European to lead an expedition into East Africa with strictly scientific motives, visited Mount Kenya. As he made his way through the dense cedar and bamboo forest, he found near the upper limit of tree growth signs of glacial striae, from which he deduced that the mountain had in the past been subjected to an intense period of glaciation. This agreed with CHAPER (1886) & DRUMMOND (1888), who had respectively found signs of an extensive glacial period in Equatorial West Africa and in Nyasaland. In the course of this expedition GREGORY made a small but significant contribution to our knowledge of the biology of the Alpine Zone. He noted among other things the abundance of rodents in the Teleki Valley; and in particular, he recognised the large number of northern plant genera that were present, and the similarity of this flora to that described from the collections of NEW (1873) on Kilimanjaro, GEDGE on Mount Elgon and STUHLMANN (1894) on Ruwenzori. He concluded that the plants represented a remnant rather than a dispersal flora. Unfortunately, most of the plants collected by GREGORY'S expedition were lost in a mishap on the Tena river during his return.

GREGORY ascended the mountain to the foot of the main peaks (16,000 ft.) and identified both their nature and the minerals which formed them (GREGORY 1896, 1921).

1896. Dr. KOLBE visited the Eastern moorland in order to collect insects, but little information about the expedition can be traced (MACKINDER 1900, JEANNEL 1950).

1899. On almost the eve of the close of the nineteenth century, in August 1899, H. J. MACKINDER approached the mountain from the West, marched through the forest to the Höhnel (Nairobi) Valley, and at noon on September 13th he became the first man to reach the summit of the highest peak, Batian (17,058 feet).

MACKINDER'S collections and observations made further contributions to our knowledge of the mountain's biology, but for Sir Frederick JACKSON'S collections on Mount Elgon in 1890, almost all the birds he found would have been new to Science. He saw and described the nest of the Scarlet Tufted Malachite Sunbird (*Nectarinia j. johnstoni*), and he brought back to Europe a skin of the Mount Kenya Hyrax (*Procavia johnstoni mackinderi*) which was described by Oldfield THOMAS (1900). Unfortunately, once more the botanical collections were dogged by misfortune when they were lost en route to Europe. MACKINDER'S notes on the geographical distribution of flowering plants were, like those made by GREGORY, of great interest.

The expedition met men of the wandering nomadic Wanderobo tribe at an altitude of 12,000 feet. This is of great interest for this tribe has now become very much reduced in numbers, and few natives now penetrate the forest in the normal course of their daily lives (MACKINDER 1900).

1908. D. E. HUTCHINS and W. MACGREGOR ROSS explored the forest and moorland, but made little contribution of biological interest (MACGREGOR ROSS 1911).

1912. CH. ALLUAUD and R. JEANNEL were the first to organise an expedition to visit the mountain with a strictly biological purpose. Their paper of 1914 on Mount Kenya was the first attempt to describe the flora and fauna as a unit, but their work was mainly entomological. JEANNEL (1950) produced later a very able general account of the mountains of Equatorial Africa.

1918. G. ST. J. ORDE-BROWNE approached the mountain from the East and described the approximate altitudinal limits of the main zones, with comments on most of the main features of the mountain's natural history (1916, 1918). It is difficult to ascertain to what altitude he scaled the mountain; the course he followed once more led to the impression that the peaks represented the side of a huge crater.

1909–1924. Between these dates the Rev. A. R. BARLOW and J. W. AUTHUR explored the mountain extensively.

1921—1922. Professor Robert E. FRIES and his brother, the late Thore C. E. FRIES, made the first major contribution to the study of vegetational zonation and taxonomy of the plants of Mount Kenya when they visited the Aberdare range and Mount Kenya between December 1921 and April 1922. Although their work on the montane forest vegetation was perhaps more detailed than that in the higher regions, they did collect to the upper limit of the flowering plants and endeavoured for the first time to delimit plant associations within the main vegetation zones (FRIES & FRIES 1948). A very large percentage of the plants collected in the Alpine Zone were new to science, and were made the subject of extensive study and publication by the Fries brothers after the expedition. One cannot speak too highly of the work done by these two scientists in bringing the study of Afro-alpine botany to the fore, and it was a sad blow to this important aspect of biological science that Thore C. E. FRIES should die so tragically from Pneumonia in Southern Rhodesia in December 1930.

1926. E. A. T. DUTTON and J. D. MELHUISH visited the mountain, climbed the peaks and explored a great deal of the peak area. They approached the mountain from Chogoria, in the East, and included in their party Lady Muriel JEX-BLAKE, who later wrote a botanical appendix to DUTTON's "Kenya Mountain" (1929). The account contains a few interesting observation, but is of little scientific value. Also included in DUTTON's work is a useful appendix by S. H. WIMBUSH, on the Forests of Mount Kenya.

In the same year J. P. CHAPIN visited the mountain, by the same route as DUTTON, and carried out useful observations in the then little known Alpine Zone (CHAPIN 1934).

1934. Carl TROLL and K. WIEN visited the mountain in 1934 where they surveyed the Lewis Glacier (TROLL & WIEN 1949). During the expedition TROLL made extensive plant collections which have recently led to his phytogeographical studies of the world's high mountains (TROLL 1958a, 1958b, 1959, 1960). It is tragic that the ravages of war should have claimed 20,000 sheets of pressed plants from the world's tropical high mountains and thus have prevented the publication of a fuller account of this invaluable comparative material. There could not surely be a clearer pointer to the importance of duplicate collecting.

1936. Colonel E. MEINERTZHAGEN collected birds extensively over the N.W. slopes above Nanyuki and thus produced the first reasonably complete list of the avi-fauna of the mountain.

1948. The Swedish East Africa expedition of this year included in their team Dr. O. HEDBERG, of the Institute of Systematic Botany, Uppsala. This was an important event for the student of Afro-Alpine Biology, for Dr. HEDBERG's work has put the flora of these regions in a new and logical order. By extensive collecting over a wide altitudinal range many of the sub-species raised by Robert and Thore FRIES were sunk into single species, while others were given new sub-specific or generic rank (HEDBERG 1951, 1952, 1954, 1955a,b,c, 1957). He is at present engaged on a comparative phytogeographical study of the Equatorial mountains (HEDBERG, 1964). From material collected on this important expedition, many groups of both the animal and plant kingdoms have been described and recorded.

1957—1958. The International Geophysical Year Expedition under the leadership of Professor I. S. LOUPEKINE studied the glaciers of Mount Kenya. Included in this expedition were B. H. BAKER (Geologist), R. JONES (Hydrologist), and the author as Biologist. The work of the expedition was largely concerned with glaciology (CHARNLEY 1960), but biological work was undertaken in a joint project with the geologist, to examine moraines in the region of present glaciers and also to study ancient low level moraines to ascertain the later stages of colonisation. It was hoped that the examination of a few moraines might yield material for dating, but this aspect of the work was unsuccessful.

1960. B. H. BAKER carried out the first geological survey of the mountain for the Mines and Geological Department of Kenya (BAKER 1961 in manuscript).

1961. Continuation of the International Geophysical Year Mount Kenya Expedition Glaciological programme, under the leadership of Professor I. S. LOUPEKINE.

1958—1961. Since the author began his studies of the ecology of Mount Kenya several workers have visited the mountain, and have made a significant contribution to our knowledge of the mountain's biology. In particular, the work of Dr. VAN ZINDEREN BAKKER of Bloemfontein, South Africa, on pollen analysis should be mentioned, and also Dr. Heinz LOEFFLER, Leader of the Vienna University Expedition to East Africa, for their work on the biology of the lakes and tarns of Mount Kenya.

PHYSIOGRAPHY

Mount Kenya is an extinct denuded volcano, that lies on the Equator some 84 miles N.N.E. of Nairobi and 270 miles from the coast (Fig. 1). The mountain is approximately 62 miles broad at its base and from a distance it gives the impression of a gentle mound rising to the sharp crags and pinnacles of the peak area. The summit, Batian, and its near neighbour, Nelion, are respectively 17,058 feet and 17,022 feet in height. The peak area represents the plug of a long extinct volcano, the present summit of which must be very much below the level of the original crater wall. GREGORY (1900) suggested that the original cone was probably as much as 23,000 feet in height.

Scattered over the mountain's rounded northern and north-

Fig. 1. Kenya General Features and Land over 10,000 Feet

eastern slopes are a number of subsidiary cones and to the east lies the largest and most impressive, Ithanguni, which rises to 11,800 feet. Another subsidiary volcano in the peak area is at the Hall Tarns, in the Gorges Valley, where all that remains is a hard plug; the crater wall has long since eroded away. This plug (14,000 feet) holds on its surface the rather beautiful Hall Tarns, in shallow glacial rock basins. To the North of the peaks two well eroded subsidiary plugs rise to 15,433 feet and 15,407 feet.

To the East and the North the mountain slopes gently down to low dry savannah country, while to the South and West it is connected to the Eastern Kenya Highlands. The Aberdare range lies to the West, and is separated from Mt. Kenya by a saddle some 50 miles wide, which at its lowest point falls to an altitude of 5,000 feet. The lower slopes of the mountain are clothed by dense montane Cedar Forest except towards the North, where the forest is replaced by Ericaceous and Protea scrub, to quite a low altitude.

Unlike the smooth little eroded cone of Kibo, on Mount Kilimanjaro 200 miles to the South-east, Mount Kenya is deeply dissected by radial valleys. These valleys radiate from the peaks and are of varying depths which, from their rounded U-shaped form, are the

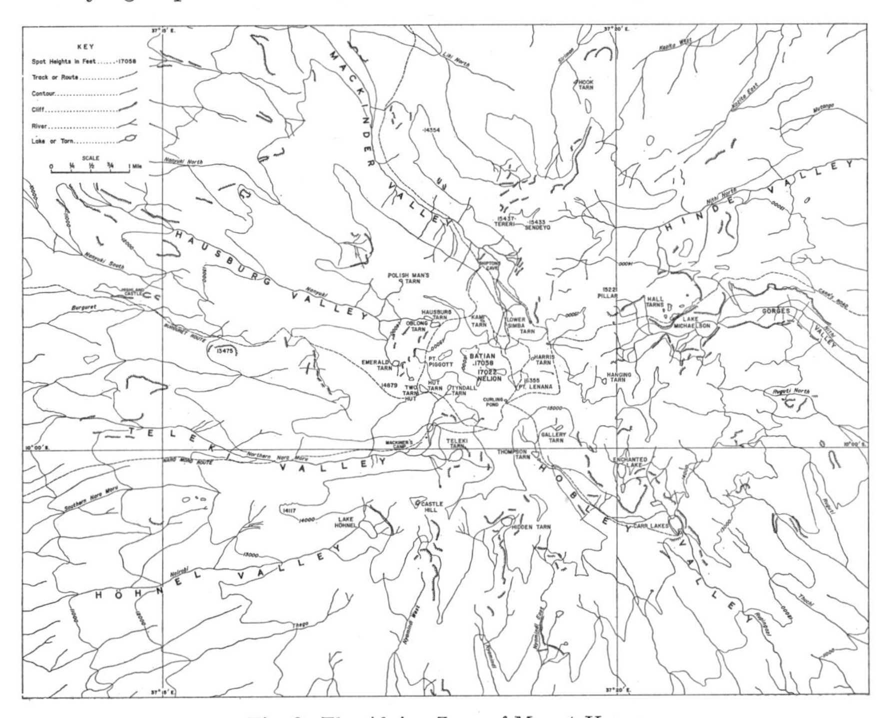

Fig. 2. The Alpine Zone of Mount Kenya

product of extensive glacial erosion. One of the largest of these is the Teleki Valley, which descends from the peaks in a gentle curve to the foot of the mountain. Passing northwards the other main valleys are the Burguret, Hausberg, Mackinder, Hinde, Gorges, Hobley and Höhnel (Fig. 2). Throughout the Alpine Zone the walls of these valleys are smooth, rounded, and often capped by rough crags and pinnacles. Around the upper edge of the forest the valleys become youthful in section and dendritic in plan (i.e. this change probably represents the approximate lowest point of the glaciers' former extension).

During the glacial period the ice erosion left a number of bowls in which lakes have since been formed. Within the Alpine Zone thirty-two such lakes exist, the most important of which may be classified as below (HUTCHINSON 1957, Vol. I):

1· Proglacial tarns

Tyndall Tarn	14,700 feet (4,480 m)
Harris Tarn	15,700 feet (4,785 m)
Kami Tarn	14,600 feet (4,450 m)

2· Glacial Rock Basins

(*a*) *Ice Scour Lakes*

Hall Tarns	14,000 feet (4,267 m)
Two Tarn	14,750 feet (4,495 m)
Nanyuki Tarn	14,700 feet (4,480 m)

(*b*) *Cirque Lakes*

Teleki Tarn	14,050 feet (4,282 m)
Lake Hohnel	13,850 feet (4,206 m)
Hausberg Tarn	14,300 feet (4,358 m)
Oblong Tarn	14,300 feet (4,358 m)
Lake Michaelson	13,050 feet (3,978 m)
Emerald Tarn	14,350 feet (4,374 m)
Hidden Tarn	14,000 feet (4,267 m)
Hanging Tarn	14,600 feet (4,450 m)
Gallery Tarn	14,650 feet (4,465 m)

(*c*) *Valley Rock Basins*

Enchanted lakes	13,850 feet (4,221 m)
Carr Lakes	13,000–13,450 feet (3962–4084 m)
Thompson's Tarns	14,100 feet (4,298 m)
Lewis Tarn	15,100 feet (4,662 m)

Many of these higher lakes are fed by ice water but those existing in cirques are largely fed by melt precipitation (often snow) percolating through scree. In the case of the Hall Tarns these small lakes exist in small depressions fed by direct rainfall and dew. The area of collection is small, being in most cases steep slopes around the Tarns, up to thirty feet above water level, hence there is considerable fluctuation in water level during the year.

Where valley walls are steep and support either free or stabilised scree, most of the water that collects filters down to streams in the valley bottoms. In these situations there is little evidence of lateral stream dissection of the slopes. Such streams only seem to occur in areas where *Senecio keniodendron-Alchemilla* scrub form stable vegetation stands and allow stable and impervious soils to form. In such regions meandering stream courses have developed and in many cases have created narrow belts of mobile scree on the valley walls.

There are several series of maps and aerial photographs that have proved invaluable during work on the mountain. They are:-

Maps.

1. 1:50,000. Kenya. GSGS 4786, Third Edition,
   Sheet South A-37 / H-1-NE
2. 1:25,000. Mount Kenya. DOS 5302, First Edition.

Aerial Photographs.

1. Royal Air Force. Mount Kenya. February 1947.
2. DOS Print Laydown. Mount Kenya. Sheet 121. 1:135,00.
3. Hunting Clan Aero-Surveys. 1GY Series.
   Teleki Valley, Peaks, Gorges Valley. January 1958.

The last mentioned aerial photographs were specially taken by Hunting Clan Aero-Surveys as their contribution to the East African I.G.Y. programme. These have been particularly useful as they were all taken from a low altitude.

From this summary of the physiography it can be seen that the Alpine Zone of Mount Kenya exists in comparative isolation from similar environments (the nearest being a small lower Alpine region on the Aberdare range, some 50 miles to the West). This region therefore forms not only an excellent situation in which to study the ecology of succession and colonisation, but as a "Terrestrial Island" it supports a flora and fauna that in comparatively recent time has been sufficiently isolated to allow distinct patterns of speciation to evolve. When more is known of the ecology of other high mountains in East Africa, there seems little doubt that much of fundamental biological significance will be discovered.

GEOLOGY

In 1893, when GREGORY visited the mountain, he reached the peak area and determined the nature of the peaks and sketched the distribution of lavas and agglomerates in the Alpine Zone. This survey remained the only true geological work done on the mountain until BAKER carried out a field survey on behalf of the Geological Survey of Kenya, from January to June 1959. (BAKER

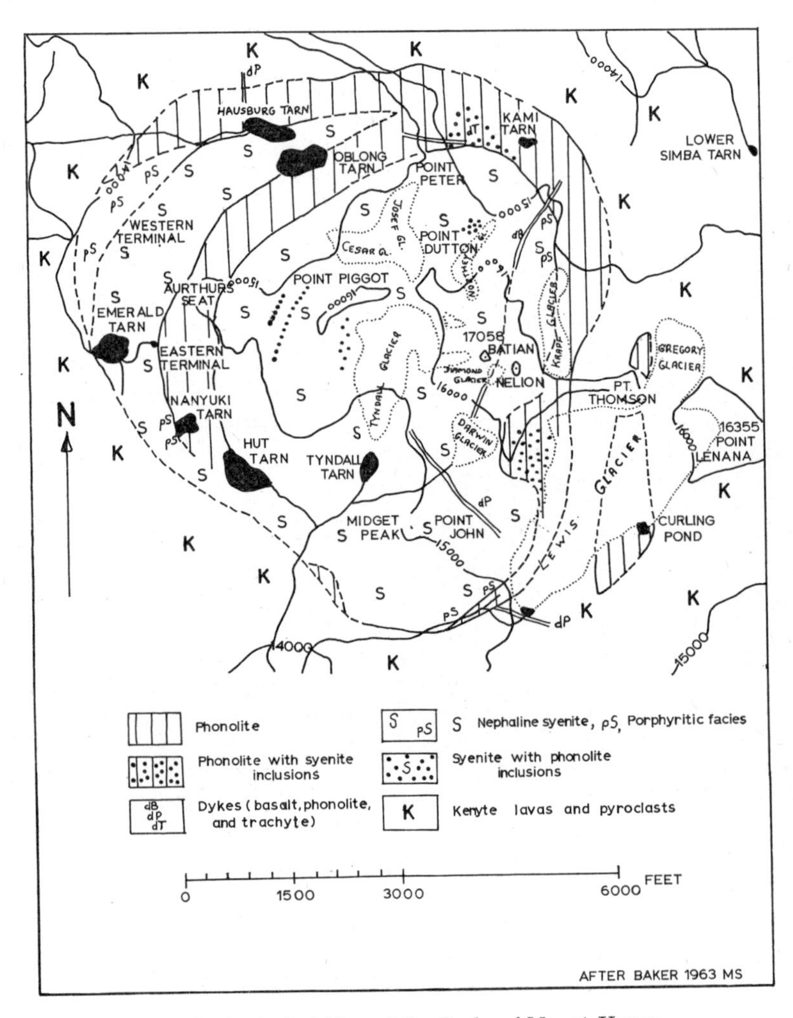

Fig. 3. Geological Map of the Peaks of Mount Kenya

1961 in MS, which he has kindly allowed the author to see and use). (Fig. 3).

The main peaks of Mount Kenya comprise the plug of a long extinct volcano which perhaps a million years ago rose as a gigantic cone to a height of about 23,000 ft.

Since this time the action of weathering and intense glacial erosion has removed all signs of the original crater wall, so that at the present time we see the denuded remains of hard plug rising from amidst extensive lava flows, the base of which is about 60 miles in diameter. A large part of the volcanic sequence is nowhere exposed, so that most of our geological knowledge is derived from the peak region, and exposures of more recent deposits along valley walls.

The rocks comprising the peaks and lavas of the slopes differ markedly in appearance although their chemical composition is very similar. These apparent differences are largely due to their relative depths when they were laid down and the consequently different rates of cooling (JENNINGS 1963).

The lavas that were in the past extruded from the main vent disposed themselves in extensive and successive flows over the mountain's slopes. These rocks are exposed on some of the minor peaks and in the lower alpine regions where the edges of individual lava flows form steep columnar jointed cliffs. All such lavas are composed of *phonolite* and their constituent rocks possess a dark grey granular base in which are embedded large (up to 2 cm) crystals of Feldspar and smaller (up to 1 cm) crystals of nepheline.

Higher up the slopes, within 3 miles of the peaks, these lavas grade into a rock with a glassy black base containing abundant crystals. This rock which is called Kenyte is a close relative of Phonolite, but attained its different form due to the fact that these lavas were rapidly cooled. Good exposures of these lavas occur near the summit of Point Lenana, but as JENNINGS op. cit. points out, materials formed near the crater rim are more often found as inclusions in agglomerates than as distinct flows.

The upper alpine zone provides many examples of the unstable terrain provided by alternate layers of agglomerate and Kenyte. This feature is particularly well illustrated on the subsidiary peaks of Tereri and Sendeyo, and the ridge top peaks of the Gorges Valley, Coryndon, Macmillan and Delamere.

The peaks of Mount Kenya are formed from the plug of this long extinct volcano. This plug is about 1½ miles in diameter and is exposed for nearly 3000 ft. of its length. Since this material was well below the surface of the mountain when volcanic activity ceased the lava cooled slowly and produced a material that is highly crystalline. These rocks are called nepheline-syenite and contain large Felspar crystals up to 3 cm long as well as smaller nepheline crystals. The plug of the main peak is surrounded by a screen of phonolite, which it will be seen in Fig. 3 is continuous in the SE-N, but divides into an inner and outer screen in the N-SW. In the South these screen phonolites are almost absent and in this situation the plug material is in contact with the more extensive lava and agglomerates.

JENNINGS op. cit. has noted that the common and closely spaced fractures found in the screen phonolites account for the large number of Tarns in the peak regions, due to the susceptibility of these regions to glacial scour action.

A number of other rock types have been found on the mountain, but their occurrence is of geological rather than ecological interest.

GLACIAL GEOLOGY

With this outline of the geology of the mountain it is important that some consideration should be given to the glacial geology, for it is glacial action that has, in large measure, shaped the mountain's surface and created the main physiographic features of the present time. In addition it is the past intensive glacial action that has to a great extent produced the soils and hence the plant associations occupying the Alpine Zone at the present time.

In his early exploration of the mountain, GREGORY (1894, 1900) realised that Mount Kenya had been submitted to a period of intense glaciation. He first came across signs of glacial action in the upper limits of the montane forest, and from a height of 11,000 feet up to the foot of the peaks the rounded valley ridges, with their occasional phonolite crags, pointed to intensive ice erosion. In fact, the general physiography led him to postulate that the mountain had in its early history been completely covered by a dome of ice, through which, at its maximum extension, only the crags and peaks cleared the surface of the ice. This is confirmed by BAKER (1959) who noted the distinct ridge top moraines that are to be observed at several points on the mountain. Fig. 4 shows the probable shape and height of the mountain before glaciation, and the glacial and fluviatile form of the Teleki Valley at different altitudes.

Following these observations, other than the minor work of MACGREGOR ROSS (1911), no glacial geological studies were made until NILSSON (1940) visited the mountain in 1927 and 1932. He considered that the large terrestrial moraines to be found in all valleys represented the maximum extension of glaciation on the mountain.

BAKER (1959) calculated that the area covered by ice at this time must have been in the region of four hundred square kilometres. He gives the following altitudes for these lower terminal moraines:

Hinde Valley (Nithi North)	11,500 ft. (3505 m)
Nithi North (Right bank tributary)	11,200 ft. (3414 m)
Gorges Valley (Nithi)	10,400 ft. (3206 m)
Ruguti North	10,700 ft. (3297 m)
Ruguti	10,200 ft. (3144 m)
Thuchi	10,600 ft. (3267 m)
Hobley Valley (Bupingazi)	10,400 ft. (3206 m)

Of great importance was NILSSON'S contention that this glaciation was a local response to the world wide pluvial and glacial epochs. (i.e. Kamasian and Gamblian pluvials and the Riss and Würm glaciations of Europe).

In the lower Gorges Valley NILSSON found what he considered

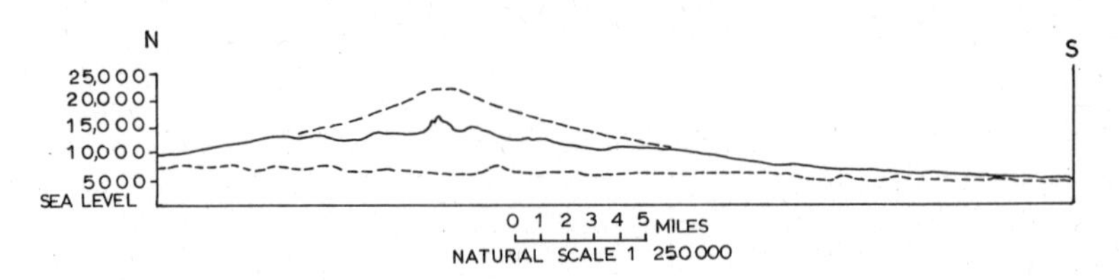

Fig. 4a. Profile showing Probable Original Shape and Height of Mount Kenya

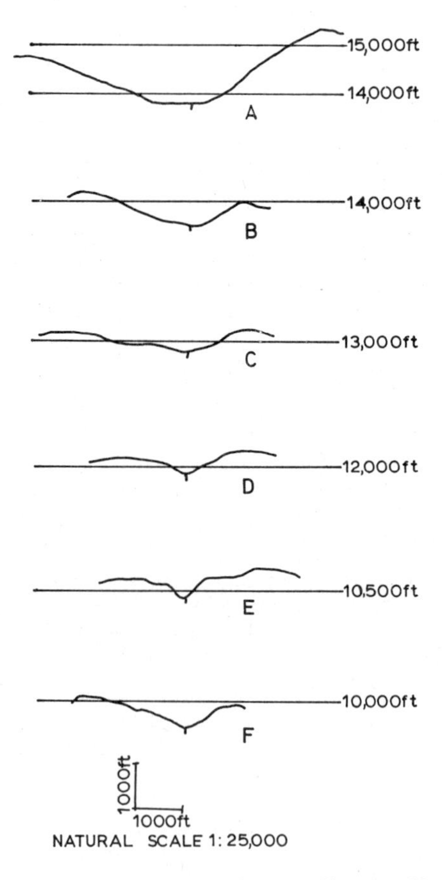

Fig. 4b. Profiles of the Teleki Valley Showing Change from Glacial to Fluviatile Form

were signs of a much more ancient glaciation underlying the more recent moraines. BAKER, in January 1958, confirmed and extended these findings in the same area, and later during his geological survey of the whole mountain (BAKER 1961 in MS.) he found further evidence on other quarters.

While these older glacial deposits are of interest to the geologist, they are of little importance to the ecologist for, except in the lower regions of the mountain (11,000 feet), they have to a great extent been obliterated by more recent glacial activity. It is by examining these later moraines that it is possible to obtain a fairly clear picture of the course of weathering of the mountain's surface, and hence to postulate in part the course of soil formation and subsequent colonisation of plants.

In most cases the moraines are still very obvious, having merely been washed clean of their fine fractions. The maximum dissection of these deposits was observed at the foot of the Gorges Valley (11,300 feet) where a gorge had been cut through a moraine to a depth of 100 feet. Naturally, in most cases lateral moraines on ridge tops are more obviously eroded than those formed on the valley floors.

At the present time glaciers on the western aspect of the peaks are much larger than those on other sides of the mountain (Fig. 5). The presence of moraines below the montane forest margin, at between 9,000 feet and 9,500 feet, shows that the ancient ice sheet must also have extended lower down on this side of the mountain. BAKER (1961 MS.) does, however, point out that one limiting factor

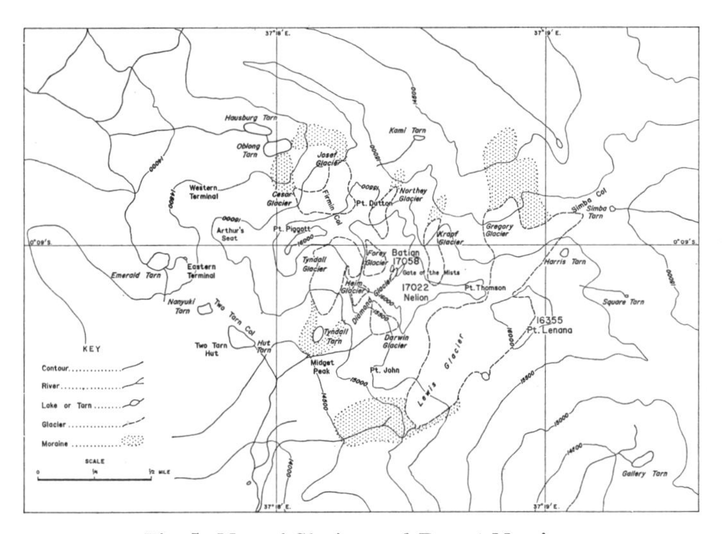

Fig. 5. Map of Glaciers and Recent Moraines

to the wide-spread extension of glaciers to the North is the shape and form of the valleys on this side.

The subsidiary cone Ithanguni (12,000 feet), in the North-east, appears to have had its own glacier system, though the degree of glaciation was not extensive (BAKER 1961 MS.).

Although there are to be found small halt stage moraines in the main valleys, the absence of any more distinct terminal moraines seems to suggest that once retreat began, it continued at a fairly steady rate over a long period. At the bases of the main peaks there are a very distinct series of terminal moraines, which form an incomplete fringe at altitudes between 14,150 feet and 15,500 feet. It is in this upper region where "historical" moraines can be observed that the primary process of habitat formation and plant colonisation can be found. These moraines show evidence of either a halt, or a local re-advance, and it will be shown that the vegetational evidence points to the latter being the more possible explanation.

From this brief summary of the glacial geology it will be seen that in comparatively recent time a large portion of the Alpine Zone of Mount Kenya has been subjected to severe glacial erosion. It is, of course, due to this coarse and violent dissection of the volcano's surface that the varied plant associations exist today from which, together with the last fragments of these glaciers and their associated moraines, we gain some small idea of the conditions under which plants and animals lived during the glacial epoch, and the pattern which early plant succession must have followed.

Plate 1. The Ericaceous Zone. Sirimon track: 10,200 feet.
The clearing in the foreground is artificial, while in the background can be seen a dense cover of *Erica arborea*, *Philippia excelsa* and *P. keniensis*, *Adenocarpus mannii* and *Anthospermum usambarense*. The main peaks and Tereri and Sendeyo can be seen in the distance.

Plate 2. Stream side vegetation in the Ericaceous Zone. 10,100 feet. The banks are thickly covered with *Erica arborea, Helichrysum kilimanjari,* and occasional specimens of the Megaphytic Groundsel, *Senecio battescombei.* The stream supports very large numbers of the larvae of the alpine *Simulium dentulosum* form *macabae.*

VEGETATION ZONES AND COMMUNITIES (Fig. 6.)

The terminology that has been used to define the vegetation zones of Mount Kenya is most varied.

FRIES (1948) considered that the Alpine Zone began above the upper limit of the Hagenia-Hypericum zone (10,829 ft., 3,300 m), while at the same time he noted that it is not always possible to draw a strict altitudinal boundary in a region with deeply cut valleys. Such undulating land surfaces produce a creeping effect of vegetation from lower altitudes, so that they extend their limits along the valley sides. This region was divided into the upper and lower alpine zone, and FRIES found the point of demarcation at approximately 12,661 ft. (3,860 m). His criterion for division was based on the distribution of the megaphytic Senecios. The lower zone was found to be occupied by the sessile rosetted *Senecio brassica,* and the upper zone by the arborescent form, *Senecio keniodendron.* These regions are said to merge along an ill-defined boundary with the Nival Zone at 14,760 ft. (4,500 m), in the immediate vicinity of the glaciers and the snow line.

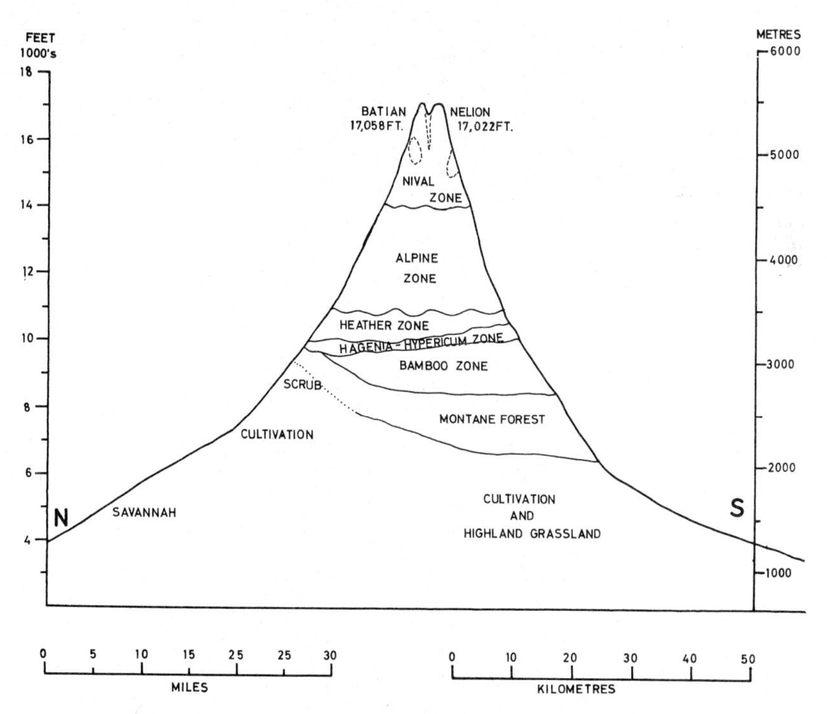

Fig. 6. The Vegetation Zones of Mount Kenya

HEDBERG (1951) accepts FRIES' classification with the important reservation that although the distribution of Senecios may be used as a method of determining the boundaries of the upper and lower alpine zones, the actual line of demarcation between these regions is not clear. A useful list of synonyms is appended to the description of the Alpine Zone in this work.

TROLL (1958), in a discussion of vegetational zonation on tropical mountains of the world, objects to the term "Alpine" owing to its close association with the winter-cold boreal high mountains, where its lower limit is demarcated by a well-defined tree line. He suggests that the term "Alpine" should be replaced by a subdivision into nival and sub-nival zones. In this connection, the sub-nival zone is synonymous with the lower and upper alpine zones of FRIES and HEDBERG. The nival zone is a very useful and important term, but where it is used in association with Equatorial mountain vegetation it is best restricted to that very narrow zone adjacent to the glaciers and recent moraines.

The argument against the use of the term "alpine" is based upon the fact that the lower limit of the alpine zone elsewhere in the world is clearly defined by a well-marked tree line. The difficulty that arises in adopting this term for Mount Kenya is the gradual thinning of the montane forest towards its upper limit into the rather diffuse, park-like, Hagenia-Hypericum zone, and this in turn gradually merges into a moorland Ericaceous zone. While there are obvious limitations to the term "alpine", bearing in mind its normal northern boreal association, the author does not consider that this term should not be applied to Mount Kenya in its broader sense of pertaining to high mountains.

The term "alpine" will here be applied to that region above the Ericaceous (moorland) zone, at 11,480 ft. (3,500 m), to the lower limit of the nival zone, 14,760 ft. (4,500 m).

It is possible therefore to say that the alpine region is arranged in a series of three altitudinal zones, bearing in mind the obvious "valley" phenomenon resulting in interdigitation of these zones, and remembering also that the local climate on different aspects of the mountain in turn exercises a profound influence on their limits in altitude.

The Alpine Zone was sub-divided by FRIES & FRIES (1923, 1948) into lower and upper regions, whose altitudinal limits were respectively 11,480 ft. (3,500 m) to 12,661 ft. (3,860 m) and 12,661 ft. (3,860 m) to 14,760 ft (4,500 m). HEDBERG (1951) accepts this classification but raises an Ericaceous (moorland) zone at the lower limit of the lower alpine zone. This sub-division presents some difficulty, since the Erica and Philippia bush, while forming a more or less continuous region above the forest, occurs in sheltered situations up to an altitude of 13,200 ft. (4,000 m). The altitudinal classification adopted by HEDBERG for the alpine zone will be used in the discussion of zonation and community structure. A brief description of the Ericaceous zone is important since these plants undergo significant transition in higher regions.

1. Ericaceous (Moorland) Zone. (Plate 1)

This zone occurs at the lower limit of the Alpine zone and above the park-like Hagenia-Hypericum zone. Due to the open nature of the upper montane forest, there is a certain amount of interdigitation between these two associations. Unlike the corresponding zone on Kilimanjaro and Ruwenzori, on Mount Kenya the Ericaceous zone forms a much less distinct belt. On the west side it is confined chiefly to the sides of ridges and along sheltered stream courses, sometimes creeping, as described above, to an altitude of 13,200 ft. (4,000 m). Passing from this aspect of the mountain, the zone widens to the North and reaches its maximum extent in the North-east above Mutindwa, and to the South-east above Embu.

The composition of this association is very variable and is not easy to define. Where pure stands occur in isolated patches (e.g. the Teleki Valley) they consist of almost closed communities of *Philippia*. This, however, cannot be taken as a general rule, for the varying climate on different aspects of the mountain has a marked effect both on the extent and on the nature of the Ericaceous zone. It will be noted later that the maximum extension of the Ericaceous zone from the North-east to the South-east coincides with the region of highest recorded rainfall above the forest (70 inches per annum). As one travels round the mountain, the closed *Philippia* forest of the West is seen to undergo a gradual change in character with the intrusion of other species. In the North-east (the Sirimon Valley) along ridge tops scattered *Protea* occurs amongst the *Philippia*, together with dense clumps of *Adenocarpus*, *Anthospermum* and *Euryops*. To the North-east (Gorges and Nithi Valleys) *Philippia* is in many places completely replaced by a dense, almost closed community of *Protea kilimandscharica*, growing to a height of four metres.

The ground flora of the Ericaceous zone varies as does that of the region with the terrain on which it occurs. Three broad habitats may be delimited:

(*a*) *Damp boggy ground:*

Characterised by large tussocks of *Festuca pilgeri* and *Carex monostachya*, the latter being less compact than the former. These enormous grass clumps have large spaces between their bases that are colonised by *Alchemilla cyclophylla* and *A. johnstonii* which, in turn, are often associated with *Geranium vagans*. From a distance these communities appear as even tracts of grassland, broken by an occasional inflorescence of *Lobelia keniensis, Kniphofia rogersii* and *Dierama pendulum*. Since the ground is sheltered by the spreading grass tussocks above, much of the ground cover is not obvious. In areas where the ground is subject to periodic water-logging, the plants tend to be restricted to the immediate vicinity of the grass tussocks. Plants to be found in this situation are *Trifolium cryptopodium, Ranunculus oreophytus* var. *oreophytus, Haplocarpha rueppellii, Swertia crassiuscula* and *S. kilimandscharica* and the creeping *Anagallis serpens*.

This habitat is particularly common in the Ericaceous zone and can be seen at the foot of every valley on the mountain. On the Naro Moru track such boggy ground is very extensive and stretches from the edge of the forest to 12,000 feet. This area is known as the "vertigal bog", since it occurs on a steep slope.

(*b*) *Raised rocky ridges or old weathered moraines:*

Where areas of bog are bordered by raised, well-drained ground, patches of the woody *Alchemilla argyrophylla* become well established, as they do in areas of seepage in the alpine zone above.

Above this narrow and by no means invariable border the true heath community becomes dominant. The three species of heath, *Erica arborea, Philippia excelsa* and *keniensis* are likely to occur together, and if they have not been burnt, the Erica may reach to a height of 12 feet. This bush is associated with slow growing bushes of *Anthospermum usambarense, Adenocarpus mannii, Protea kilimandscharica* and *Euryops brownei*. Of smaller stature in this association are *Struthiola thomsonii, Helichrysum chionoides* and *Habenstretia dentata*. Scattered in small clearings or around the larger shrubs are a wide variety of herbs. These include the small attractive *Erica whyteana* sp. *princeana, Helichrysum kilimanjari, Satureja pseudosimensis, Silene burchellii* and *Wahlenbergia aberdarica*. The ground supports a rich growth of moss and the heaths are often "bearded" with *Usnea* in such situations as these.

(*c*) *Stream courses:* (Plate 2)

In the Ericaceous zone the streams have often cut steep gorges in the bottoms of valleys, thus forming protected niches wherein plants may grow to profusion. As the walls slope down to the water, the heaths reach a maximum height of over twenty feet. These plants are associated with the megaphytic *Senecio battescombei* and robust stands of *Helichrysum kilimanjari*. The ground flora is extremely rich and contains both forest and alpine representatives. Along the shaded courses of the streams, tussocks of *Carex bequertii* and *Juncus capitatus* are commonly to be found, and between these plants the rich flora includes *Alepidea masaica, Disa stairsii,*

Gladiolus watsonioides, Delphinium macrocentrum, Senecio subsessilis, Gerbera piloselloides, Dicrocephala alpina, Crepis carbonaria and *Trifolium burchellianum.* Creeping over areas of bare ground and rocks, *Sedum ruwenzoriense, Galium glaciale* and *Geranium simense* are common, and where the streams pass through deep gorges, clumps of bamboo, *Arundinaria alpina,* occur at over a thousand feet beyond their normal range.

One of the commonest features of the upper moorland are the low, stunted bushes of *Erica* and *Philippia,* consisting of thin, fairly young shoots arising from a stout woody basal stem. HEDBERG (1951) suggests that this is probably due to fire pressure. He cites previous records by MACKINDER (1900), FRIES (1923) and SYNGE (1937), and his own personal observations in 1948 of extensive burning, both here and in the alpine zone. The author has also observed the effects of fire, in the Lower Gorges and Sirimon valleys; in both situations the upper limit of moorland vegetation is definitely suppressed by fire, Erica existing in the Lower Gorges as only small stunted bushes, up to one metre in height, whose thin shoots rise from blackened basal stumps. In January 1958 extensive fire damage was observed over a wide area, with the Erica and Philippia seeming to be more affected than Protea, which is also common in the area. Again in July 1959 there was evidence of fire damage earlier in the year, and the plants had just begun to regenerate. The same observations were made in the Sirimon valley in August 1961. In this case the fire had been very recent and was, without doubt, due to human agency. Here the only plants that had started regeneration were Festuca tussocks, while the Heath showed no signs of growth. When this area was revisited in December 1962 the Erica stumps had still not revived and were apparently quite dead.

Undoubtedly the fire hazard in this zone is considerable and can be largely attributed to human agency, though lightning may be of secondary importance. Moreover large areas of the moorland were burnt during the Mau Mau Emergency, both by security troops and by terrorists. It would nevertheless be too speculative to suggest that the moorland zone would extend very much higher without the intervention of fire, for in areas of the mountain which are seldom visited by man the heath does not reach appreciably higher limits.

2. The Alpine Zone.

This region shows an important and interesting vegetational transition with increase in altitude, culminating in a narrow nival zone. The upper limit of flowering plants is marked by scattered and isolated occurrences of *Helichrysum brownei* on the main peaks, up to 16,000 feet and probably higher. Lichens occur on the summit of Batian itself, at 17,058 feet.

The interdigitation of the moorland and alpine region has already been mentioned. Owing to the deeply dissected nature of the slopes of Mount Kenya, it is only possible to speak of altitudinal zonation of the alpine region in the broadest terms, and it will be seen that community structure and the physical nature of their habitats are the main criteria to be borne in mind when considering this region.

One may only see really well demarcated zones on more recent mountains, such as Kibo, the main crater of Kilimanjaro, where the effect of glaciation has been negligible and valley phenomena are almost absent from the consideration of vegetational zonation. For instance, the saddle between Mawenzi and Kibo is covered from about 14,000 feet upwards with an almost even tussock Helichrysum scrub community, which is interrupted only when rocky masses produce micro niches that have been colonised by other flowering plants.

Fries (1948) used the vertical distribution of *Senecio keniodendron* and *S. brassica* to divide the upper and lower alpine zones respectively. Although it was admitted that both species occurred sporadically in the other zone, Fries considered this criterion to be of sufficient importance to use the presence or absence of these plants as a major indication of zone. Hedberg (1951) accepts this classification with the reservation that there is considerable overlapping in the distribution of *Senecio brassica* and *S. keniodendron*. *Senecio brassica* certainly occurs low down in the alpine zone, and since it grows as a sessile rosette it is very distinct from the arborescent *S. keniodendron*. The latter does not begin until about 12,661 feet (3,860 m) which seems a reasonable marker for determining the lower limit of the upper alpine zone. *S. brassica*, however, does not stop at this point and in the region of Emerald Tarn (Housberg Valley) its upper limit reaches within 500 feet of *S. keniodendron*. One possible point of confusion in the lower alpine zone is the presence of another megaphytic groundsel, *S. battescombei*, which occurs throughout the moorland and lower alpine zones, mainly along stream courses. This species is thin-stemmed, and tends to branch more than the stoutly built *S. keniodendron*; it is most readily distinguished by its inflorescence and its leaves.

It is not possible to over-emphasise the fact that while for the sake of description it is convenient to separate the alpine zone into an upper and lower region, such a division is artificial and should be used with caution in vegetational studies. Both plant and animal community studies seem to suffer from the danger of over-simplification; and for this reason the alpine zone will be treated in this work from the point of view of a series of plant associations existing within the broader and more artificial framework of an upper and a lower alpine zone.

Excluding the ubiquitous species of tussock grass, the most

obvious plants occurring in the alpine zone are species of the genera *Senecio, Lobelia, Alchemilla* and *Helichrysum.* Since these plants are important community indicators, they will be considered both as individual species and in relation to their adaptation to their environment, before the actual communities in which they occur are described.

Senecio LIN. (Compositae)

HEDBERG (1957) treats the taxonomy of these plants in some detail; in particular, he points out that intra-specific variability is considerable. This accounts for the large number of species raised by FRIES on variability, which were later condensed by HEDBERG after he had examined a wider altitudinal range of material. There is little doubt that the effect of climate on altitudinal variation is considerable. Even at the same altitude factors of position and aspect have a profound effect on growth form and morphological differentiation.

Subgenus *Dendrosenecio* HAUMAN.

The group contains the tree-like megaphytic species with woody or underground creeping stems (COTTON 1944).

Senecio brassica R. E. FR. and TH. FR. JR. (Plate 3)

Occurs as a low rosette with a dense tomentum on the backs of the leaves. Creeping woody stem, covered in dead vegetable matter, from which the rosettes arise in the axils of leaf scars. Inflorescence erect, up to 3 feet in height, with bright yellow flowers. They appear to flower en masse at irregular intervals. These plants are the first of the large Senecio to occur in the alpine zone from 10,829 feet (3,300 m) to 14,109 feet (4,300 m). In suitable situations they form dense ground cover, on damp ground with standing water. They bear many fine roots, but the system is neither deep nor extensive.

Endemic on Mount Kenya.

Senecio battescombei R. E. FR. and TH. FR. JR.

A thin-stemmed woody plant up to 18 feet in height. Branches more distinct than *S. keniodendron.* Inflorescences up to four feet with bright yellow flowers not unlike those of *S. brassica*; occurs in the lower alpine zone close to stream courses, or on boggy ground from 9,512 feet (2,900 m) to 12,464 feet (3,800 m). Scattered specimens occur above this level in suitable locations. They overlap with the lower limit of *S. keniodendron* and may be confused with this species when not in flower, although the situations in which they are found are quite different.

Endemic on the Aberdare range and on Mount Kenya.

Senecio keniodendron R.E. FR. and TH. FR. JR. (Plate 4)

Rather more robust than *S. battescombei* and more sparsely branched due to suppression of growing points. Up to 18 feet in height. Inflorescences with dull purple bracts, lacking ray florets; occurs from 11,500 feet (3,500 m) to 15,300 feet (4,665 m). In the middle of its altitudinal range it forms patches of almost closed forest in association with *Alchemilla*.

The most continuous *Senecio* forest occurs in the Mackinder Valley, where the walls are almost completely covered from 12,500 feet (3,811 m) to 14,000 feet (4,268 m) (Dec. 1962). The densest stands are found on sloping ground where surface water is readily available. They do not do well on ground which contains standing water. The distribution of this plant clearly indicates that the main factor controlling the occurrence of *S. brassica* and *S. keniodendron* is a habitat rather than conditions related to altitude. Close to the peaks the ground becomes steep and will not support damp boggy ground, thus excluding *S. brassica* from entry into the plant associations at higher levels. However, where suitable patches of ground do occur (e.g. at the bead of the Hausberg and Gorges Valleys, at 14,000—14,300 feet) the plants do flower quite normally.

Endemic on Mount Kenya. (One doubtful record from the Aberdare range. HEDBERG 1957).

Subgenus *Eusenecio* O. HOFFM.

Herbs and shrubs of more normal size and stature.

Senecio roseiflorus R. E. FR.

A small slender herb, 3 to 4 feet in height. Flowers pale mauve; leaves sticky and covered with glandular hairs. Occurs in shaded situations in the moorland zone and in the lower alpine zone, where it is mainly found on raised rocky ground from 9,512 feet (2,900 m) to 12,628 feet (3,850 m).

Endemic on Mount Kenya and on the Aberdare range.

Senecio purtschelleri ENGL.

Perennial herb, up to 2½ feet tall; a stout plant branching from the base. Usually on raised rocky ground, with a preference for rock shade increasing with altitude, from 10,997 feet (3,350 m) to 16,400 feet (5,000 m).

Endemic on Mount Kenya, Kilimanjaro and Mount Meru.

Senecio keniophytum R. E. FR.

A low herb of great importance in the upper alpine zone. This plant is remarkable in its range of variation of pigmentation, growth form and pubescence. In sheltered situations on the mid-alpine zone the plants may reach .8 feet in height, with sparse hairs, while

Plate 3. *Senecio brassica,* the almost sessile "giant" Groundsel: 13,000 feet. Note the persistent leaf frill around the base of the plants, and the white tomentum on the underside of the leaves. This lining is often stripped to line the nests of the Scarlet Tufted Malachite Sunbird.

Plate 4. *Senecio keniodendron,* the erect megaphytic groundsel of the Alpine Zone: 14,000 feet.
Note the very large persistent leaf frill, also specimens of S. *brassica* in the foreground.

Plate 5A. A small "family group" of *Lobelia keniensis* on the side of Lake Höhnel: 13,800 feet
The inflorescence in the background is just beginning to elongate. Note in this species that the bracts are broad and short, thus exposing the flowers, also that the broad leaves of the rosettes hold water in which a small species of Chironomid breeds.

Plate 5B. *Lobelia keniensis* rosettes and *Alchemilla johnstonii*. The rosettes are connected by underground stems, and contain small pools of water in which colonies of Chironomids breed.

Plate 6A. On a moraine at the head of the Teleki Valley the sunlight catches the long hairy bracts of *Lobelia telekii*. This area is typical of the dry well-drained ground on which this plant usually occurs: 14,000 feet.

Plate 6B. An elongating inflorescence of *Lobelia telekii* on well drained ground in the Sirimon Valley. A small rosette is seen maturing at the base. Note the long hairy bracts of this "Ostrich plume" Lobelia which completely obscure the flowers. In the background a mature inflorescence can be seen on the right.

in the upper alpine and nival zones, as exposure increases, there is a distinct tendency to produce a low mat form profusely covered with hairs. A detailed study of the altitudinal variations occurring in this plant would prove rewarding. Range from 11,972 feet (3,650 m) to 15,908 feet (4,850 m). It appears to be a secondary coloniser at lower altitudes and a primary coloniser in the nival zone.
Endemic on Mount Kenya.

Senecio aequinoctialis R. E. Fr.

Medium herb, branching basally. Occurs from 9,840 feet (3,000 m) to 13,940 feet (4,250 m). Since these plants are not common, they are of little importance in community studies in the alpine zone.
Endemic on Mount Kenya and the Aberdare range, although insufficient material is available for study. (Hedberg 1957, p. 242 and p. 362).

Lobelia L. (Campanulaceae)

In the alpine zone of Mount Kenya all species of this genus occur as low compact rosettes, from which rise tall inflorescences up to 6 feet in height. Inflorescences and creeping stems are hollow and produce a copious white sticky latex. A clear gelatinous material is produced by the leaf bases of the rosettes which mingles with the rain-water that collects in between the flowers to form a viscous fluid. This secretion may in some way lower the freezing point of the water and in turn protect the leaves against sub-zero temperatures at night.

Lobelia keniensis R. E. Fr. and Th. Fr. Jr. (Plate 5)

A megaphytic plant with a hollow stem that occurs as broad rosettes, rarely singly, more commonly in "family" groups. Broad glabrous leaves. Rosette a tall hollow spike, with deep purple flowers, well exposed. Inflorescences up to 5 feet in height. The first Lobelia to occur in the moorland and alpine zones are found in moist, semi-bog adjacent to lakes and streams. Occurs sporadically at 10,000 feet (3,050 m) and reaches a maximum density in the mid-alpine zone. Found in sheltered situations up to 14,268 feet (4,350 m).
Endemic on Mount Kenya.

The rosettes of these species are of particular interest for they arise from a branching underground stem out of the axils of the leaf scars, in much the same way as those of *S. brassica*. It is of interest to compare these alpine species with those that occur in the montane forest. These are *L. gibberoa* and *L. bambuseta*, both of which have erect stems up to 20 feet in height. No doubt in this thick forest habitat it is an advantage for the leaves to be borne on erect stems, and it seems possible that the sessile alpine species have been derived from some such erect forest form. Should such an erect plant occur in the alpine zone, it would not survive long without the protection of a thick bark, such as is to be found in the megaphytic Senecio. The creeping stem of *L. keniensis* and single stem stock of *L. telekii* are protected from the cold frosty nights by being buried in the ground. Although these underground stems are quite shallow, even at a depth of four inches the temperature remains remarkably constant and seldom, if ever, sinks to freezing point.

Lobelia telekii SCHWEINF. (Plate 6).

Megaphytic, more distinctly monocarpic than *L. keniensis*, though in suitable positions "family groups" are not uncommon. The rosette is much less robust than that of *L. keniensis*; leaves are narrow with a distinct waxy bloom. Inflorescence clothed with hairy bracts, flowers are hidden, reduced and almost colourless; up to 6 feet in height. Usually to be found on well-drained, coarse grained material, but may exist in damp areas when this condition has arisen after the plant has become established (Sirimon valley, August 1961). Although *L. telekii* has been observed here and there at a low level, 10,168 feet (3,100m) to 10,486 feet (3,200m), it does not occur in any appreciable frequency until about 11,480 feet (3,500 m). Specimens have been found in the nival zone at 15,088 feet (4,600 m) (Tyndal Glacier December 1957, January 1960, Lewis Glacier January 1961, Northey Glacier 1963). Since no signs either of fresh or dead inflorescences could be found, it must be assumed that these plants are at the limit of their range and are sterile.

Endemic on Mount Kenya, the Aberdare range and Mount Elgon.

Alchemilla L. (Rosaceae)

Stout shrubs to creeping procumbent herbs. These plants form a distinct Alchemilletum in the alpine zone, in some places existing in pure closed stands on well drained ground. Other species form dense ground cover in damp grass tussock areas.

Alchemilla cyclophylla TH. FR. JR.

A prostrate creeping herb, occurring in damp situations from the Ericaceous belt into the Upper Alpine Zone. Not common in natural communities, but it seems to colonise both man-made and animal tracks. It forms a dense ground cover in tussock grassland from the upper forest clearings to 13,940 feet (4,250 m) and is found colonising bare ground with mosses. Only described from Mount Kenya and the Aberdare range.

Alchemilla argyrophylla OLIV. in HOOK.

Low stout shrub, with many branched stems. It has a wide

distribution throughout the alpine zone from the Ericaceous belt to the upper alpine zone, 15,252 feet (4,650 m), generally on well-drained slopes and forming pure closed stands in suitable localities. Stems clothed with red-brown scale leaves. Hairiness seems to be a variable character, those plants growing in the protective shade of rock outcrops and old moraines being less hairy than those in exposed situations. May reach up to 4 feet in height.

Endemic on Mount Kenya, the Aberdare range and Kilimanjaro.

Alchemilla johnstonii OLIV. in HOOK.

Small shrub, with prostrate or procumbent stems. Branches frequently and may produce erect stems up to 4 feet in height. Common on damp, semi-boggy ground, at altitudes up to 14,760 feet (4,500 m), and usually associated with tussock grassland and *S. keniodendron* where the latter occurs on valley walls with surface water. Like *A. argyrophylla,* it forms an almost pure Alchemilletum in suitable situations (e.g. stream sides).

Endemic in tropical East Africa.

Helichrysum MILL. (Compositae)

Nearly always stout hardy shrubs. Growth form and other morphological features such as leaf shape and hairiness vary greatly with altitude and aspect.

Helichrysum odoratissimum L.

Suffrutescent thin-stemmed plants with small yellow flowers. Common in lower alpine zone in a variety of habitats, from open tussock grassland to rock outcrops and old moraines. Not of great importance in alpine communities. Up to 2½ feet in height.

Widespread over Africa from sea level to alpine grassland.

Helichrysum cymosum (L.) LESS ssp. *fruticosum* (FORSK) HEDB.

Perennial woody shrub, up to 3 feet tall, with dirty white flowers. Occurs largely on raised well-drained ground. Small weak plants found as high as 16,400 feet (5,000 m) in the Teleki Valley.

Common over most of Africa except in the South.

Helichrysum citrispinum DELILE var. *armatum* (MATTF.) HEDB.

Much branched, spinous shrub, with white flowers; up to 2 feet in height and may reach nearly 3 feet in sheltered situations. Common on raised rocky ground throughout the alpine zone.

Endemic on Elgon and Mount Kenya.

Helichrysum brownei S. MOORE.

Hardy shrub forming low compact bushes up to 1½ feet tall. Flowers white to cream. May occur on damp patches on steep ground, seems able to grow in rock cracks with very little soil. Throughout the alpine belt to the main peaks where the author has found plants growing from cracks on steep rock faces at 16,000 feet.

Endemic on Mount Kenya and the Aberdare range.

Helichrysum chionoides PHILIPSON.

Tall woody herb up to 6 feet tall. Sparse branches arising from the top of the rigid stem. Flowers medium size, white to fawn. Occurs through the alpine zone up to 12,628 feet (3,850 m). Seems to prefer well drained grassland, raised rocky outcrops and steep stream banks.

Endemic on Mount Kenya and the Aberdare range.

Helichrysum nandense S. MOORE.

Endemic in East and Central Africa.

Helichrysum meyeri-johannis ENGL.

Endemic on certain mountains in East Africa.

Helichrysum ellipticifolium MOESER.

Endemic in Kenya and one record from the Congo.

Helichrysum formosissimum SCH. BIP. ex A. RICH. var. *formosissimum*.

Widespread over Congo, East Africa and S. Ethiopia.

Helichrysum guilemii ENGL.

Widespread over the Congo, East Africa and S. Ethiopia.

These last five species of *Helichrysum* all occur on Mount Kenya, but the distribution is sporadic and they are not important in community structure.

While this is in no way a phytogeographical study, it is of interest to notice the high percentage of endemism amongst these more obvious representatives of the alpine flora. Of the twenty-one species cited above, five are endemic on Mount Kenya, nine on Mount Kenya and the Aberdare range, and the other species are endemic on Mount Kenya and Kilimanjaro, Mount Kenya and Mount Elgon, or fall into the class of seven species which exhibit wider distribution. It is of interest to note that endemism seems to be a more common feature amongst the genera *Senecio*, *Lobelia* and *Alchemilla* than amongst the *Helichrysum* species. Perhaps this reflects the greater degree of success of the former in colonising the alpine zone.

All the genera cited above are very obvious components of the alpine zone, with the megaphytic Senecios and Alchemilla scrub forming almost pure stands under suitable conditions.

In studying the ecology of this zone one is particularly struck by the great variability of all these plants. HEDBERG (1957) has studied this phenomenon and as a result he has lumped together many species that appear to have been raised in the past from either a single or a small number of specimens. The variation of growth form with increase in altitude is particularly striking. In addition, the variations that may occur within a single altitudinal zone owing to factors of shade and aspect, though perhaps less obvious, are very significant.

It is not possible to consider this subject at great length, but certain plants enumerated above are worthy of note. In an environment where the vegetation is subject to great heat and radiation during the day, and intense cold at night, one would expect to find morphological variation that will enable the plant to cope with a reasonable degree of success in these extremes.

The variability in all these four genera is considerable, HEDBERG (1957) includes them in his extensive and important appendix to his study of the taxonomy of the Afro-Alpine Flora. In many cases where a whole series has been examined and old species have been sunk into a single species, it is evident that previous workers have described species on the simple basis of altitudinal variation. Nearly all the plants in the alpine zone possess protective features to one degree or another.

SALT (1954) has pointed out that plants living at high altitude are not adapted simply to cold. It will be shown, when consideration is given to the Alpine climate, that the plants in these regions need to survive not only cold, but also intense heat and radiation. In addition, perhaps the greatest need of the plant is to be protected not against one of these extremes, but rather from the great rapidity with which conditions change from one to the other.

Perhaps one of the most noticeable adaptations to high altitude conditions in all parts of the world is the rosette habit. On Mount Kenya we may consider the alpine vegetation to be divided into three growth layers:

1. The aerial layer occupied by the megaphytic *Senecios*,
2. The herb layer occupied by *Helichrysum* and grass tussocks,
3. The ground layer occupied by a profusion of low rosette, procumbent, or creeping plant species.

This latter category is occupied by *Haplocarpha, Ranunculus, Haplosciadium, Swertia, Anagallis, Veronica, Romulea* and *Carduus*.

Plants living in the alpine zone must be protected from high radiation (re-radiation) and the pronounced diurnal climatic regime.

The high percentage of species sharing a rosette habit in the alpine zone might seem to be a protection against low air temperatures. When, however, one considers (p. 58) the fact that the degree of temperature change at ground level is considerably higher than that of the air this cannot be the most direct explanation. It is true that many of these plants have either hairy or thick shiny cuticles that probably protect them from excess radiation. Any plant though with an erect stem in these regions is protected by either a persistent leaf frill or a thick corky cortex, for a stem projecting into an atmosphere that may fall as low as -10 °C must be in considerable danger of freezing unless it is surrounded by some means of insulation.

Although the top soil is submitted to nightly frost heaving, temperatures measured from 2 to 4 inches below the surface remain remarkably constant, day and night at 1-4 °C. Thus although a root is not frozen in these soils the rate at which water can be drawn up into the stem must be very slow. Under these circumstances there seems little doubt that there would be a considerable advantage in developing a very short stem, for not only is the distance that water must be transported to the leaves reduced, but the danger of freezing is also considerably lowered. Indeed almost all plants possessing such a rosette habit also have short stout tap roots.

Another factor that does not appear to have been considered previously, in relation to the development of the rosette habit in these regions is atmospheric humidity. During field work on the mountain the author has found that the humidity of the air alters violently from +90% early in the morning to below 20% in bright unobscured sunlight. The humidity, however, at ground level remains remarkably constant and seldom falls below 70%, except on bare, stony ground. Thus the leaves of a rosette plant would lie within the sphere of influence of this high and more constant humidity and transpiration and consequent water loss would be greatly reduced.

It is extremely difficult to separate modifications that may protect a plant against excessive radiation from those that may be of assistance in the realm of temperature insulation. Certainly the very common hairy or tomentose coverings found on *Alchemilla argyrophylla, Helichrysum chionoides, H. cymosum, H. citrispinum* and *Senecio keniophytum* may well reflect a great deal of incoming radiation while at the same time protecting the plant at night against sub-zero temperatures. It can surely be no coincidence that leaves bearing these coverings are always folded on the growing parts. This feature is demonstrated in all the species mentioned above.

The insulating function of hairs may be twofold, for where a layer of relatively immobile air is trapped above the leaf surface, transpiration losses will be considerably reduced.

An adaptive trend that may be almost entirely related to protection against radiation is the common occurrence of a red pigmented epidermis, as is to be observed in *Lobelia keniensis, L. telekii* and

Swertia volkensii. From field observations it appears that *Senecio keniophytum* not only shows increased hairiness with altitude, but also deepening of epidermal pigmentation.

Most of the *Helichrysum* species to be found in the alpine zone show consistently white or off-white capitulae. SALT (1954) suggested that these surfaces might well be a great protection by reflecting a large percentage of incoming radiation. The driver of a motor car in the tropics is only too well aware of the remarkable effect of painting the roof of a car white, for this action reduces the temperature inside by several degrees. At the time of writing, experiments are being conducted with thermistors to measure the degree to which these or other such modifications can protect the plant.

Any advantage that is gained during the day by reflection from these bright capitulae is extended at night when the involucral bracts are folded tightly into "night buds".

The short erect reduced leaves of *Erica* and *Philippia* which are sparsely hirsute also afford the stem a degree of protection from radiation and low temperature, though these two genera possess a layer of bark that must in itself provide some degree of insulation. In the same way the hairy leaf bases of *Helichrysum cymosum* and *H. chionoides* also provide an efficient covering for a large percentage of the stem's surface.

The grass tussocks of the alpine zone which cover by far the larger part of the region's surface achieve their protection in another way. Most of these grassy masses appear, except when flowering, to be dead. In many cases the centre of the tussock is completely rotten, leaving a complete or incomplete ring of living material. When a large tussock is sectioned it is found that dead stems rise to over a metre in height from a thick basal disc. This platform holds the main growing points, and in addition has a good water holding capacity and a high humus content.

The effectiveness of their insulating capacity is demonstrated by temperatures measured during the night in the Mackinder Valley. While the air temperature and that between the outer leaves of a tussock was —5 °C, that in between the basal leaves, and hence in the region of the growing points, was +1.5 °C. It appears that the gap between freezing and not freezing is no more than a couple of degrees, just as in the hot springs of Lake Magadi *Tilapia grahami* regularly feeds in water that is within 2 °C of the lethal temperature (COE, 1966).

After fires in the tussock grassland it has been noted (COE, January 1963, Field Observations) that mass flowering of tussocks takes place shortly afterwards. While production of seed is important for dispersal, this new growth must at the same time be vital in restoring the insulating layer that is so essential to the survival of the basal growing points.

The megaphytic Senecios and Lobelias also face problems of radiation and temperature change, and their adaptations to these features of a high altitude existence are perhaps some of the most striking of all. More so because although many authors have commented on their peculiar growth forms little experimental work has ever been done to elucidate the manner in which they survive these extremes.

Lobelia keniensis, L. telekii and *Senecio brassica* all preserve a rosette habit although many are up to 3 feet in diameter. Each of these three species has its own individual manner of protecting itself against the rigours of the alpine climate.

All these species in their vegetative phase produce enormous rosettes from which at a later date an inflorescence will be thrust several feet into the air. These rosettes all show remarkable sleep movements, for at night their leaves close into a tight night bud to re-open the following morning (HEDBERG 1963, COE, Personal Field Observations). Although the rosettes open and close in this manner the central growing point of the rosette is always tightly ensheathed by young leaves which will remain folded until the inflorescence is formed. Temperatures measured in all these rosettes by the author showed that even when the air temperature fell to —5 °C the temperature amongst the outer leaves was —1 °C, while the central leafy cylinder remained up to 2 °C above freezing throughout the night.

Both the *Lobelia* species possess thick cuticles and varying degrees of reddish pigmentation, which probably protect the leaves against radiation during the day when they are open. *Senecio brassica* shows an additional feature for it possesses a thick tomentum on the underside of its leaves, which the sunbird *Nectarinia j. johnstoni* uses to line its nest. During the day this covering may seem to be little protection to the rosette, but at night when the rosettes are closed it must provide a considerable degree of protection from reradiation.

The rosette leaves of *Lobelia keniensis* are broader than those of *L. telekii* and form a small reservoir in which water collects. This liquid is found in all the rosettes even when there has been very little rain, in spite of the low humidity of the atmosphere during the very common periods of unobscured sunlight. The liquid is thick and mucilagenous, with little resemblance to rain water, but rather a fluid that is in part at least produced by the thick white fleshy leaf bases. A great feature of interest in these reservoirs is that no matter how low the temperature falls at night the liquid only freezes on the top with a thin layer of ice, towards the outside and inside, thus producing a wedge of ice that is open at the bottom and hollow inside. If, as seems likely, the inclusion of this secretion lowers the freezing point, here again we observe

Plate 7. Interdigitating stands of *Senecio keniodendron* (dark) and *Senecio brassica* (light), on the wall of the Teleki Valley: 13,800 feet.

Plate 8. *Senecio keniodendron – Alchemilla argyrophylla* forest on the wall of the Mackinder Valley: 13,700 feet.
Note the ring-like formation of the *Alchemilla* in the foreground, where it has died off in the centre of the clump.

another unique measure to protect the plant from low night temperatures. A small Dipteran (*Nematocera,* Chironomidae) has utilised this micro-environment to lay its eggs and at night when the surface freezes their temporarily moribund larvae lie in the unfrozen water below the ice. Their very presence and the fact that they occur in every rosette seems to bring evidence that these reservoirs are never frozen solid. By comparison with these small quantities of water in a rosette, a pint can of water left outside a tent is frozen solid by midnight.

The closing of the rosettes appears to be related to the degree by which the temperature of the atmosphere is lowered, for it has been observed by the author that on cloudy nights when the temperature may fall just below 0 °C the degree of closing of *Lobelia* and *Senecio* rosettes was much less than on unobscured nights with temperatures as low as —10 °C.

When rosettes give rise to inflorescences the degree of protection that can be afforded to the developing flowers is not as clear as it is in the rosettes. All these inflorescences when unfolding at the tip are either protected as in the two species of *Lobelia* by bracts or in the case of *S. brassica* by long hairs. When, however, the flowers mature they are more susceptible to low night temperatures although they are protected from radiation by their bracts. *Lobelia telekii* has long hairy bracts (ostrich plume lobelia) completely obscuring the flowers, which are in consequence almost colourless. By contrast the bracts of *L. keniensis* are broad and pointed and the flowers are visible. The vital parts are hence enclosed by the bract bases and temperatures of +1 °C were recorded at the base of a corolla at 8 p.m. when the air temperature was —4.5 °C (Mackinder Valley, January 1963, Personal Field Observations).

The inflorescence of *Senecio brassica* has its flowers protected by lax hairs until they open, but after this the only protection they have is the rather thick bases to individual bracts. No temperatures have yet been recorded in flower heads, but these are shortly to be carried out with thermistor probes.

The Megaphytic *Senecio keniodendron* being an erect plant (whose average height is 6-8 ft.) shows further adaptations to the alpine climate. Discounting the inflorescences of the two alpine species of *Lobelia* and *S. brassica* this plant towers many feet above any other plant in the alpine zone.

The leaves of their terminal rosettes, which open and close like *S. brassica* and the *Lobelia* sp. are dark green and covered with a thick cuticle. They do not possess the tomentum as in *S. brassica.*

Temperatures measured inside a rosette of this species showed close agreement with those recorded for other rosettes, remaining between +1 and +2 °C throughout the night.

A plant whose stem rises several feet above the ground requires additional protection to ensure that the fluids in its stem are not frozen. *S. keniodendron* achieves this by means of its persistent leaf girdle, for when the leaves die they remain attached to the plant for a very long time. A plant 8 feet high was covered with dead leaves down to a level of 1 foot above the ground, and below this the main stem was covered by a thick layer of deeply ridged cork.

The average diameter of a *Senecio keniodendron* stem is 6½ inches while the cork layer is up to 10 inches and the leafy girdle up to 1′10″. Thus on any one radius the stem proper is protected by 1.75 inches of cork and an additional 6 inches of leaf girdle. The insulating capacity of this layer of poor conductor and entrapped air is evident, but temperature measurements demonstrate it in a more striking fashion. The temperature at the bases of the leaf frills remains almost constant day and night at +4°C with a variation of little more than one degree.

It is not surprising that in an environment where shelter for animals is so scarce these leafy girdles are used as roosting places by the three resident passerine birds, *Nectarinia j. johnstoni, Pinarochroa sordida earnesti,* and *Serinus s. striolatus.* This microhabitat is in addition used by beetles (Carabidæ and Curculionidæ), spiders (Lycosidæ) and molluscs.

The Teleki Valley has become regularly used by climbers in the last eight years, and in so doing man has had a secondary effect on the formerly dense stands of *S. keniodendron.* Since fire wood is difficult to come by, the leaf girdles are stripped off by visitors and in so doing they deprive these plants of their insulating layers. The large number of dead "stripped stems" in this region compared with other little visited areas of the mountain leaves little doubt that the removal of the leafy girdles is sufficient to kill these plants. Unlike Mount Elgon, which has had native cattle grazing on it for many years, Mount Kenya is thus only just beginning to feel the influence of man as an ecological factor.

3. The Lower Alpine Zone.

This zone is characterised largely by open tussock grassland with scattered groups of *Lobelia keniensis* rosettes. Such open grassland occurs chiefly on gentle sloping ground. The soil consists of a black top soil with a high humus content underlaid by dark red clay. This material has been largely deposited through particles that have been washed down from ridges and valley walls.

These areas being close to the Forest and Ericaceous belt are frequently crossed by well-defined Elephant "roads". One such track which crosses the lower Naro Moru valley is well marked, and appears to allow animals to pass from one side of the mountain to another without crossing the steep water-worn valleys of the montane forest. Since the upper edge of the montane forest marks the upper end of the V-shaped pluvial valleys, such passage is more easily accomplished across the lower ends of the U-shaped glacial valleys of the lower alpine zone.

There is a pronounced absence of shrub-like plants in this zone, and the main field layer is occupied by tussocks of *Festuca pilgeri* ssp. *pilgeri*. These large grass clumps grow to a height of over three feet. The edges of the tussocks produce new grass shoots, while the centres are usually dead and rotten. Their stout damp bases afford an ideal position for mosses, and other small grasses, to grow, thus producing a massive tussock circled at the base by plants that have colonised this sheltered niche. At higher levels flowering is sporadic, but in regions where there is fire hazard, mass flowering of these grasses is often seen, producing a strong contrast with the brown, unburnt and non-flowering tussocks around them (Gorges Valley January 1958, July 1959, Sirimon Valley August 1961).

Small streams often cut their way through the top and sub-soil to expose the glacial gravel base. The sides of these streams and the outer edges of its plant associations, where it grades into rocky ridges, support small groups of *Senecio brassica* and *Lobelia keniensis*.

Between the tussocks of *Festuca* less robust tussocks of *Agrostis trachyphylla* occur, which often seem to get their first hold in the community at the bases of the former. They increase in frequency towards the upper alpine zone and on recently colonised moraines these tussocks form the primary grass coloniser of the area.

Wherever the ground rises to rocky lava crags, or moraine deposits, the coarser-leaved *Deschampia flexuosa* becomes common. This association extends along ridge tops to an altitude of 15,000 feet, giving them a characteristic pale fawn-yellow colour, which can be distinguished from some distance and which forms very distinct patches on aerial photographs.

Semi-boggy patches in this region support rich moss growths and small groups of *Lobelia keniensis*. *L. telekii* is found at the upper limit of the zone on well drained ground. Between the tussocks, communities of creeping rosette or sessile plants occur. In general it can be said that this ground will only be colonised where it receives some measure of protection from the field layer. In patches where, due to animal or frost agency, the soil is disturbed, colonisation by flowering plants is difficult until the soil has first been stabilised by mosses and lichens. Predominant among these low plants, which in places form a dense ground cover, are *Ranunculus oreophytus* var. *oreophytus*, *Anemone thomsonii*, *Haplocarpha rueppellii*, *Alchemilla johnstonii*, *Geranium simense* and *Swertia crassiuscula*.

Since the inclusion or exclusion of a plant from an association is a function of the habitat they occupy, the vegetation of this region will be described by reference to the sites on which they occur.

(*a*) *Flat or gently sloping ground — usually wet.*

These sites occur at the bases of all valleys in the alpine zone (Northern Naru Moru Valley 11,500 to 12,000 feet, Gorges Valley 11,500 to 12,000 feet, Sirimon Valley 11,000 to 12,000 feet).

This ground may be described as damp, semi-bog, bearing an almost continuous cover of *Festuca pilgeri* ssp. *pilgeri* and *Festuca abyssinica,* both of which are often over three feet in height, and small tussocks of *Carex monostachya.* Scattered over this grassland are small groups of *Lobelia keniensis,* which usually indicate seepage or a certain amount of standing water. The rosettes of this plant stand about one foot high and are about 18″ in diameter. These compact rosettes hold small pools of water in which a small Chironomid larva lives. They form an important item of diet of the Scarlet Tufted Malachite Sunbird (*Nectarinia johnstoni johnstoni*). Where this damp ground borders raised rocky areas, tussocks of *Deschampsia flexuosa* and *Anthoxanthum nivale* occur together with small groups of *Senecio brassica.*

Due to the spreading nature of these massive tussocks, bare patches of dark, humus laden soil occur between them. These areas are colonised by mosses and a number of rosette plants that include *Haplocarpha rueppellii, Swertia crassiuscula, Alchemilla johnstonii, Ranunculus oreophytus* var. *oreophytus, Geranium simense, Haplosciadium abyssinicum,* and small clumps of the stout *Lycopodium saururus. Bartsia petitiana, Veronica glandulosa* and *Viola emenii* occur sporadically but are not common.

(*b*) *Weathered and eroded ridge tops.*

This is a fairly constant habitat in the alpine zone, beginning in the lower alpine zone as isolated phonolitic outcrops and extending into the upper alpine zone. Here they become more rugged until at about 14,000 feet, owing to the devastating effect of frost, they are entirely devoid of vegetation. A typical example of this well drained habitat occurs from 11,500 feet, in the Teleki Valley. In some areas a similar habitat has been created by extensive weathering and erosion of lateral and ridge top moraines, such as those existing in the lower Gorges Valley at 11,750 to 12,000 feet. It is interesting that though these two areas are very similar in the type of substrate that has been created, due to the difference in aspect their vegetation is quite distinct.

Ridge tops on all sides of the mountain are rounded and sweep down to U-shaped glacial valleys. These ridges bear pinnacles and crags of phonolitic and basalitic material that provide both good drainage and protection to very distinct plant associations. The more open sections are covered with grass tussocks of *Deschampsia flexuosa* and *Pentaschistis minor*. The stature of these plants decreases as the degree of exposure of the ridge increases. Where shelter is offered by rocky outcrops the number of woody plants increases rapidly, so that in places dense clumps of *Philippia excelsa, Helichrysum chionoides, H. citrispinum, H. cymosum* and *Alchemilla argyrophylla* are to be found where the degree of shelter is high. Under rocky masses thick growths of moss, *Senecio purtschelleri, S. roseiflorus* and *Arabis alpina* are a distinct feature of these ridges on all quarters of the mountain.

The surface between tussocks and other plants is not so well colonised by rosette plants as is the damp ground in the same region, due largely to unimpeded drainage and frost heaving. The commonest plants occupying these situations are *Galium ruwenzoriense, Veronica glandulosa* and *Swertia crassiuscula* which, owing to their creeping habit, seem to be very successful on these comparatively mobile surfaces.

Close to and amongst large rocky outcrops, which are usually occupied by colonies of the Mount Kenya Hyrax (*Procavia johnstoni mackinderi*), small communities of a quite different association occur. These small pockets of vegetation are shaded and protected from the sudden temperature changes of the more exposed situations. Here are found remarkably green and lush stands of the aromatic *Heracleum elgonense,* up to 3 feet in height, sometimes accompanied by the robust thistle *Carduus keniensis* and *Arabis alpina.*

In the vicinity of Hyrax burrows rocky ledges often collect masses of dung that has been washed down or deposited by these animals who, like the Porcupine, regularly defaecate in the same spot. These faecal deposits support large numbers of small molluscs and dense masses of the bright yellow *Sedum ruwenzoriense.* It is interesting to note that these plants are seldom, if ever, cropped by the Hyrax and it would seem more than fortuitous that they so often camouflage the holes that lead into the colony. The association may well have arisen by the gradual elimination of these plants that are regularly eaten by the Hyrax, to the advantage of those that are not taken as a regular item of the animals' diet, the resulting association being of mutual advantage and leading to the survival both of the plant and the Hyrax. At higher levels, 13,500 to 14,500 feet, *Senecio purtschelleri* becomes a common plant around Hyrax colonies and is also seldom eaten.

Although with increase in altitude the frequency of occurrence of different plant species may change, by and large the association remains remarkably constant until it almost entirely dies out between 15,000 and 15,500 feet. In this situation the criterion of the distribution of the megaphytic Senecios is not a fit one for distinguishing a lower and upper alpine zone. Scattered *Senecio keniodendron* plants occur on well drained ground leading from these outcrops from 11,500 feet to 15,500 feet, i.e. covering almost the maximum range of *Senecio brassica.*

A number of very large terminal and lateral moraines occur at the lower reaches of most major valleys. The most distinct groups of these were visited at the foot of the Gorges valley, where a series of concentric moraines lie on either side of the Nithi River. These deposits stretch from about 11,500 feet, at the foot of the valley, to about 12,500 feet where they merge with the main northern ridge of the Gorges Valley.

It may be said that in many ways the vegetation of these ridges is distinctly Ericaceous, since the range of these plants is considerably extended owing to the high rainfall on this part of the mountain.

In the region of these moraines two distinct associations are found: one being the moraine ridges themselves which consist chiefly of glacial boulders and gravel, while the other is the distinct strips of grassland that grow on the finer material that lies between them.

On the moraine deposits the main impression given is that of a woody moorland community limited to the boulder strewn moraine tops. Shrubs growing here include *Adenocarpus mannii, Anthospermum usambarense, Euryops brownei, Helichrysum chionoides, Phillippia excelsa, Protea kilimandscharica* and *Struthiola thomsonii.* Both the moraine boulders and the shrubs present a great deal of shade where more delicate plants can become established. Prominent among the grasses are small tussocks of *Deschampsia flexuosa* var. *flexuosa* and *Pentaschistis minor.* Other plants represented are *Bartsia kilimandscharica, Dierama pendulum, Gladiolus watsonioides, Disa stairsii,* and *Scabiosa columbaria.*

Between the moraines lie strips of grassland that have become established on the fine moraine fraction that has long ago been washed out of the ridges. Due to the protection afforded to this grassland by the moraine ridges, and the fact that a high clay fraction in the soil keeps it continuously damp, a number of plants occur here somewhat above their normal range.

Tussock grasses make up the most obvious ground cover in this habitat (± 90% of the ground cover) and are made up of *Festuca pilgeri* ssp. *pilgeri, F. abyssinica, Agrostis trachyphylla* with *Anthoxanthum nivale* occurring along the edges. All the herbs that occur do so on the patches of bare ground between the tussocks. The following were recorded in January 1958: *Erica whyteana* ssp. *princeana, Galium glaciale, Geranium kilimandscharicum, Geranium vagans, Ranunculus keniensis, Satureja pseudosimensis, Selago thompsonii, Silene burchellii, Swertia crassiuscula, Trifolium cryptopodium* var. *kilimandscharicum,* and *Wahlenbergia aberdarica.*

Such strips of inter-moraine grassland slope gently away at their lower ends to the river banks where the soil is more or loss waterlogged. Here the number of species occuring between the tussocks increases considerably. These include *Alepidea masaica, Anagallis serpens* ssp. *meyeri-johannnis, Ardisiandra wettsteinii, Carex bequertii, Crepis carbonaria, Dicrocephala alpina, Dipsacus pinnatifidus, Gerbera piloselloides, Juncus capitatus, Sabaea brachyphylla, Senecio battescombei, Senecio subsessilis, S. brassica, Swertia kilimandscharica, Trifolium burchellianum* var. *johnstonii, Utricularia afromontana, Viola emenii, Veronica abyssinica.* It will be noted that this association is similar to that described along stream sides in the Ericaceous zone. This may be explained by the considerable degree of protection from cold valley winds which this area enjoys, and also by the higher rainfall on this side of the mountain. In fact, almost all these herbs are not found above the forest line in the Teleki Valley.

4. The Upper Alpine Zone.

This zone can be demarcated by the presence of the megaphytic groundsel *Senecio keniodendron* occurring in distinct communities between 12,500 feet (3,811 m) and 15,000 feet (4,573 m). At both ends of this range they are sparse; they reach their maximum development between 13,000 feet (3,963 m) and 14,000 feet (4,268 m). Since these plants are used as zone indicators, it is useful to begin a description of this region with the habitat in which they are most common.

(*a*) *Valley Walls.*

Throughout the upper alpine zone the glacially eroded valleys bear along their sides distinctive vegetational communities. It is in part the stature of the plants of these associations that makes them appear so distinct, the ground cover being in fact fairly typical for damp, well-drained ground throughout the whole alpine region.

One of the most characteristic vegetational features of these valleys is the interleaving tongues of *Senecio keniodendron* that reach down to the valley floors, and the reciprocal tongues of *Senecio brassica* that climb up the valley sides between those of their megaphytic counterpart (Plate 7). This can be explained largely by the nature of the surface they are occupying. It has already been stated that small streams pass down the valley walls through *S. keniodendron* forest. These streams when they reach the valley floor have deposited a fan of fine material that, due to the colloidal nature of these deposits, remains for the most part waterlogged, and it is upon these surfaces that *S. brassica* grows and thus ascends the valley wall.

Along the walls of all major valleys the most noticeable feature of plant cover is the presence of dense stands of *Senecio keniodendron*, the bases of which are surrounded by a thick interwoven mass of *Alchemilla argyrophylla* (Plate 8). This association forms a nearly continuous undulating band that runs on along the valley side, on a fairly shallow but humus laden soil and a good supply of sub-surface water. Small streams usually meander down the valley walls through these stands.

In situations where these plants form a dense community they do so to the almost complete exclusion of other plants, *Agrostis trachyphylla* being the only plant capable of entering the association and surviving. Towards the valley ridges and also towards their bottoms, however, the density of these plants diminishes and other plants succeed in entering the association. Undoubtedly the entry of *Ranunculus oreophytus, Arabis alpina* and *Haplosciadium abyssinicum* depends on a good supply of surface water and shade. Towards the ridge tops where drainage is more rapid and most of the soil available is frost heaved every night, the *Alchemilla* becomes stunted and is replaced as a primary field cover by *Agrostis trachyphylla, Deschampsia flexuosa* and *Anthoxanthum nivale,* which together form dense tussocks over 3 feet in height. The ground between these tussocks is colonised by *Swertia kilimandscharica* and *Valeriana kilimandscharica* ssp. *kilimandscharica.* The tussocks here are often found with *Galium ruwenzoriense* climbing over them.

In the upper montane forest, *Arundinaria alpina,* in preference to a continuous zone, exists largely on steep slopes. This slope factor is also evident in the upper alpine zone. The steep walls of the Teleki valley contrast with the completely different community to be found at a similar altitude (13,000 feet to 13,500 feet) in the Gorges Valley. In this situation the presence of the Hall Tarns plug provides a vertical cliff face on which little can survive except on weathered ledges. At the foot of the cliff steep masses of debris grade from rocky boulders against the cliff itself, through a decreasing gradation to mere particle size on the edge of Lake Michaelson. Apart from producing a fairly unique habitat, the protection afforded by these high walls has allowed the vegetation of the lower alpine zone and moorland to extend its range considerably beyond its normal upper limit. The same phenomenon is true of plants that have crept from the montane forest into the Nithi Gorge (11,200 feet), also in the Gorges Valley.

On the ledges of the Hall Tarns cliff well established groups of *Gladiolus watsonioides, Sedum ruwenzoriense* and *Anemone thomsonii* can be found at almost 14,000 feet.

Below the cliff the boulder scree supports a rich and well protected flora that includes *Anthoxanthum nivale, Arabis alpina, Blaeria filago, Deschampsia flexuosa, Erica arborea, Habenstretia dentata, Helichrysum citrispinum* var. *armatum, Lobelia telekii, Philippia excelsa, Satureja punctata,* and *Senecio keniophytum.*

Between the boulder scree and the edge of lake Michaelson an equally rich flora is found, protected by scattered boulders and large tussocks of *Deschampsia flexuosa, Festuca pilgeri* ssp. *pilgeri, F. abyssinica* and *Agrostis trachyphylla*. These are *Alchemilla johnstonii, Carduus keniensis, C. platyphyllus, Geranium kilimandscharicum, Haplocarpha rueppellii, Haplosciadium abyssinicum, Luzula abyssinica* ssp. *aequinoctialis, Senecio brassica, Swertia volkensii, Valeriana kilimandscharica* ssp. *kilimandscharica* and *Wahlenbergia pusilla*.

Around the lake shore *S. keniodendron* and *S. brassica* form some of the most continuous stands to be seen on the mountain. Boggy patches are particularly common around the shore and where the bases of *Agrostis trachyphylla* tussocks are immersed in water, lush growths of moss, liverwort, *Ranunculus oreophytus, Subularia monticola, Crassula granvikii* and *Pentaschistis minor* are all well established.

(*b*) *Valley Floors.*

The ultimate effect of the daily cycle of freezing and thawing which is so typical of the alpine zone of the Equatorial mountains is particularly evident in valley bottoms. These habitats owe their existence to soil flow following continuous frost heaving and to glacio-fluvial outwash. Where valley walls are steep these habitats merge into one another, the latter often being covered by the former. Once more, however, it is true to say that there tends to be a certain amount of overlap of plants with a wide tolerance of surface conditions.

The basic constituents of both communities are tussock grasses which are in many situations (Teleki Valley, 13,750 feet, Gorges Valley, 14,100 feet) cropped low by Hyrax. These tussocks include *Agrostis trachyphylla* on damper ground and *Festuca pilgeri* ssp. *pilgeri* and *F. abyssinica* on drier ground. *Anthoxanthum nivale* forms a border to the associations where the ground becomes rocky and what little soil there is, is submitted to frost heaving.

(*i*) Glacio-Fluvial areas at valley heads. (Teleki Valley 13,500 feet to 14,500 feet)

This particular site was close to the large moraines at the head of the Teleki Valley frequented by large colonies of Hyrax. The grass tussocks of *Agrostis trachyphylla* and *Pentaschistis minor* were in consequence eaten almost to ground level (6 inches). The area was characterised by low plants such as the very attractive *Myosotis keniensis, Nannoseris schimperi, Cerastium afromontanum* and *Carduus platyphyllus*. At the edge of the association bordering the moraine were thick mats of *Valeriana kilimandscharica* ssp. *kilimandscharica*, which occurs in damp sheltered situations from here to the top of all valley ridges. Possibly due to good drainage the area was noticeably free of bare ground, i.e. frost heaving was noticeably free of bare ground, i.e. frost heaving was negligible. This association is common in other valleys also and has been seen at the heads of the Mackinder, Gorges and Hausberg valleys at similar altitudes.

(*ii*) Solifluction terraces and other damp, almost flat ground adjacent to streams. (Teleki Valley 13,500 to 14,500 feet)

In these regions the ground is fairly evenly covered with tussocks of *Festuca pilgeri* ssp. *pilgeri*, which do not grow as large as they do at lower altitudes. *Agrostis sclerophylla* and *Carex monostachya* also occur in the association on waterlogged ground. The area between the tussocks is well colonised by mosses, *Alchemilla johnstonii, Galium glaciale, Swertia volkensii* and *Romulea keniensis*. Damper patches of ground close to the streams support small, almost pure clumps of *Luzula abyssinica, Senecio brassica, Lobelia keniensis. Cardamine obliqua, Oreophyton falcatum, Peucedanum friesiorum, Subularia monticola, Crassula granvikii, Haplocarpha rueppellii* and *Ranunculus oreophytus* occur in profusion on damp ground adjacent to streams, or at areas of seepage at the bases of valley walls.

Above 14,000 feet the percentage of damp ground falls off sharply owing to the absence of appreciable areas of flat ground and more particularly to the previous nature of the terrain. The effect of this porosity is clearly seen at the head of the Teleki Valley (14,500 feet), where streams from the Tyndal and Lewis glaciers flow within ten feet of one another until they converge at 14,250 feet. The former supports a rich vegetation along its banks, while the latter is almost completely devoid of vegetable growth. This may be explained by examining the terrain over which the water flows on its way to its point of confluence in the valley bottom. The stream from the Tyndal glacier flows almost entirely out of sight, hidden beneath boulder scree, while that from the Lewis glacier is exposed throughout its course from the glacier to the point where it runs close beside the stream from the Tyndal glacier. When these streams were examined at 10 a.m. it was found that the one which proceeds from the Tyndal glacier was flowing freely, while that from the Lewis glacier was covered with a layer of ice half an inch think. The apparently minor difference in the terrain over which these streams flow, although it may lead to only a few degrees difference in water temperature, is quite sufficient to encourage, or completely to suppress, plant growth.

(*c*) *Ridge Tops.*

The nature of such surfaces depends in large part upon the degree of slope of the valleys that flank them. Such ridges have in many cases been free of ice action for a much longer period than the valley bottoms lower down, and have in consequence lost most of their soil cover. The vegetation that occupies such a position must of necessity be hardy and able to withstand both the effect of wind and the extensive frost heaving that takes place in this exposed habitat.

Nearly all the smoother, unbroken ridges are capped by sparse tussock cover. This vegetation occurs up to about 14,500 feet, above this altitude the ridges become more rugged and the vegetation is almost entirely absent at 15,000 feet. It has already been pointed out that this community can in many ways be regarded as an extension of that found in the same situation in the lower alpine zone.

Below the brow of these tussock covered ridges *Alchemilla argyrophylla* forms what may be considered as almost the only closed stands of vegetation in the upper alpine zone. The only other plants to enter such associations are the two high level species of woody everlastings, *Helichrysum cymosum* ssp. *fruticosum* and *H. citrispinum* var. *armatum*. The existence of this dense Alchemilletum is undoubtedly controlled to a large extent by the degree of slope. It has already been shown that on the steep walls of the Teleki Valley *Senecio keniodendron* occurs with *Alchemilla* as an under-cover. The presence of an almost pure Alchemilletum is characterised by a gentle slope from a ridge, usually descending into one of the smaller subsidiary valleys.

The intrusion of rocky outcrops on these ridges breaks the vegetational sequence by the introduction of *Sedum ruwenzoriense, Heracleum elgonense, Arabis alpina, Valeriana kilimandscharica* ssp. *kilimandscharica, Senecio keniophytum* and *S. purtschelleri* wherever a small degree of shade is provided. *Helichrysum brownei* can be found growing out of cracks in these rocks and occurs from here up to the limit of flowering plant growth at 16,000 feet on the main peaks.

Sedum crassularia occurs on small patches of coarse gravel, while *Galium ruwenzoriense* and *Veronica gunae* often colonise the edges of scree slopes. Rocky crags that support small patches of soil support scattered tussocks of *Koeleria convoluta*. These marginal areas will be mentioned later for they play a minor but interesting role in the colonisation of the higher regions of the upper alpine zone.

(*d*) *Lakes and Tarns.*

The retreating ice sheet of the alpine zone left in its wake a number of Tarns and Lakes in almost every valley on the mountain. These areas of more or less permanent water have created a niche that has been successfully occupied by distinctive plant communities. Lakes occur throughout the alpine zone from Lake Michaelson (13,000 feet) to Two Tarn (14,500 feet) and above this, proglacial tarns occur at the feet of the glaciers. While it will be shown that the aquatic vegetation is almost uniform, the plant associations of the shore line vary with altitude, although in many cases this variation is probably attributable to the degree of exposure of the surrounds.

Most of the main Lakes and Tarns were visited, but only those will be described which, by virtue of their altitude or aspect, show an important feature of lakeside vegetation.

Lake Höhnel (Nairobi Valley 13,700 feet):

This lake lies in a cwm at the head of the Nairobi Valley, and is surrounded on three sides by steep cliffs with a dam at its lower end formed by a large terminal moraine. It is not, however, entirely a moraine dammed lake, for this material overlays a dense rocky barrier behind which a rock basin has been scoured away by glacial action.

The most obvious plants associated with this and other lakes are the small succulent *Subularia monticola* and *Crassula granvikii,* both plants being found either on wet banks or completely submerged. They appear to be able to flower equally well in either situation. *Subularia monticola,* when found submerged, usually grows in small pools or channels close to or connected to the main lake; the main criterion for their occupation here seems to be a preference for a gravelly bottom and water that is little disturbed by wind or inflowing streams, whereas *Crassula granvikii* occurs chiefly on the soft diatomaceous ooze that aggregates on the gravel shelf around the shore line. In this position it creeps over the ooze, where it frequently forms striking patterns on the surface, where it frequently forms striking patterns on the surface.

Lake Höhnel has a rather solid gravel bottom, shelving near the shore and reaching a deep central area. This region has been colonised by a rich and profuse growth of *Potamogeton.* Neither HEDBERG (1957) nor other workers have found evidence of this plant flowering and one must therefore conclude that it is probably sterile. It is of interest that this is the first record of this genus from the alpine zone of an East African mountain. Although no flowers were observed, DANDY (in HEDBERG 1957) considers that it probably belongs to *P. schweinfurthii* A. BENN.

The Lake is surrounded by flat boggy ground, which exists as a platform between the water and the steep cliffs at the head of the valley. The plant association occurring in this situation is very similar in composition to that of damp ground throughout the alpine zone. The bases of the large tussocks which are always water logged are circled by rich growths of *Alchemilla johnstonii, Haplocarpha rueppellii* and *Ranunculus oreophytus.* Between the tussocks, areas of damp black soil support clumps of *Luzula abyssinica, Haplo-*

sciadium abyssinicum, Montia fontana ssp. *fontana,* and small plants of *Lobelia keniensis* which, since they were not connected to rhizomes or underground stems of other plants, were probably seeded.

Below, the lake shore rises rapidly to a steep terminal moraine and along the shore line on this side of the lake a narrow belt of the community enumerated above is to be found. Where the well-weathered moraine rocks start to rise, a sudden change in the constitution takes place. The tussock grasses in this area are large and well spaced; scattered amongst them small groups of *Senecio brassica, S. keniodendron,* and rosettes of *Lobelia telekii* are common on this well drained ground, which also supports rich growths of *Cerastium afromontanum* var. *afromontanum, Nannoseris schimperi, Carduus platyphyllus, Valeriana kilimandscharica* ssp. *kilimand-scharica* and *Helichrysum citrispinum* var. *armatum.* Masses of dead *Alchemilla argyrophylla* occur all along this ridge. This phenomenon has been observed in similar situations on the mountain, and it seems clear that this is due to the gradual removal of the fine fractions from these ridges until the point is reached when woody plants are unable to survive. Where the Alchemilletum thins out suddenly towards the top of valley walls, it seems likely that the same phenomenon may be operating there also.

Teleki Tarn (Teleki Valley 14,100 feet):

This large Tarn occurs in a steep-sided cwm at the head of the valley. Due to the very steep walls, a large area of scree exists around the edge of the lake which, owing to its continuous movement, has not been colonised. Around the lake shore there are spasmodic areas of vegetation, where damp rocks have been colonised first by lichens and mosses and later by flowering plants (Fig. 7). The lake itself has the "poached egg" pattern of so many of the alpine lakes. The reason for this deep central position is particularly obvious on Lake Teleki when the nature of the ledge around the shore has been examined. The continual washing of particles into the lake from the surrounding scree slopes has created an uneven ledge of gravel and detritus, which is wider and more extensive on the sides of the lake adjacent to the steeper and more mobile scree. This platform bears a layer of diatomaceous ooze, several feet thick on the Eastern side.

The lake is similar to Lake Höhnel with *Subularia monticola* in sheltered inlets and *Crassula granvikii* growing in great profusion on the ooze platform. It is this latter species that is responsible for the remarkable "fairy ring" patterns of this lake. The same feature is found in other lakes, but it is nowhere quite as obvious as on Lake Teleki. These rings form a crescent around one side of the lake and occur as either complete, incomplete or fused ring structures. When examined closely the outer edge of the ring is found to be

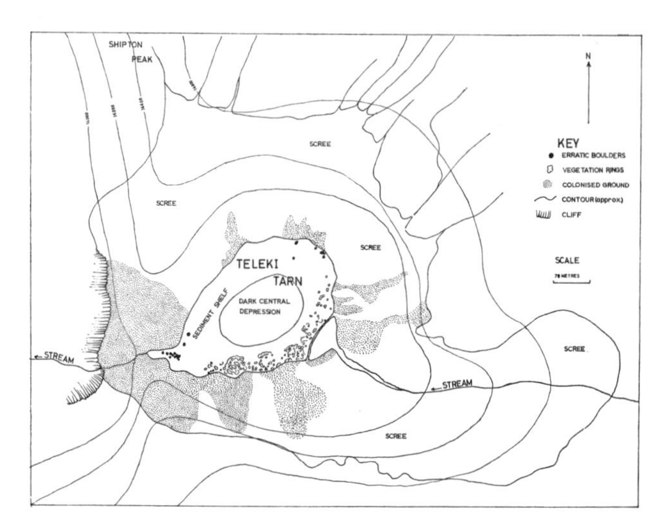

Fig. 7. Cwm containing the Teleki Tarn.

bright green in colour, grading to a dirty brown on the inside. On examination it was found that this was due to regeneration of a creeping stem stock to the outside, while inside the ring the plants were dying off. All the plants in the centre of the ring were covered with diatoms, in many cases the whole plants being completely encased in them. Both this fact and the slow settling out of dead planktonic organisms is responsible for the death of the older plants.

The scree slopes show a graduation of particles in size from the top to the bottom. Around the shore line the finer fraction has been washed out and all that now remains is a coarse gravel, in which little evidence of frost heaving could be found. These portions of the shore line and lower scree slopes that have been colonised support widely separated grass tussocks of *Agrostis trachyphylla*, together with odd plants of *Anthoxanthum nivale* and *Deschampsia flexuosa* var. *afromontana* higher up the scree slopes.

Senecio keniodendron occurs around the lake shore, particularly along stream courses flowing into the lake. Above the shore line the ground cover consists of *Nannoseris schimperi*, *Haplosciadium abyssinicum*, *Valeriana kilimandscharica* ssp. *kilimandscharica* and *Carduus platyphyllus*. The *Ranunculus oreophytus* and *Haplocarpha rueppellii* of lake shores at lower altitudes are absent; this is probably due as much to the lack of a humus rich surface soil as to altitude.

Emerald Tarn (Hausburg Valley 14,250 feet):

The lake is dammed by a moraine and rock fall, and is surrounded by steep cliffs. Again the lake has a deep central portion, surrounded by an ooze covered ledge. *Subularia monticola* and *Crassula granvikii* are both present in the water, the latter forming a single ring around the central depression rather than the "fairy rings" seen in Lake Teleki.

Although at nearly the same altitude as the Teleki Tarn it will be seen that the flora associated with the shore of Emerald Tarn is very much richer. This may be explained by the absence of cold glacier winds, that affect the former, and to the presence of quite a large area of bog associated with the latter.

Very obvious amongst the vegetation around the Tarn are large groups of *Senecio brassica* sheltered below cliffs on the Northern side. The presence of these Giant Groundsels at 14,200 feet once more demonstrates that, given the right conditions of shade and aspect, most alpine associations can and do exist at almost all altitudes, although large communities are limited to certain zones. Associated with these plants on slightly raised, well-drained slopes are tussocks of *Festuca pilgeri* ssp. *pilgeri, Deschampsia flexuosa* var. *afromontana* and occasional plants of *Koeleria convoluta*. Between these tussocks the low rosettes or creeping forms associated with these areas include *Cerastium afromontanum, Haplosciadium abyssinicum, Carduus platyphyllus, Valeriana kilimandscharica* ssp. *kilimandscharica* and *Senecio keniophytum.*

On the South-eastern corner of the Tarn a large area of boggy ground supports a typical bog association. Tussock grasses include *Festuca abyssinica* and *Agrostis trachyphylla*. Dense groups of *Carex monostachya* occur around the tussocks, or as secondary growth in the centres where the grass has started to die back. The bases of these tussocks support lush growths of moss, *Ranunculus oreophytus* and *Haplocarpha rueppellii,* which also colonise bare ground between the grasses. Around the edges of the bog *Swertia volkensii* and *Romulea keniensis* enter the association.

Hall Tarns (Gorges Valley 14,100 feet):

These series of small tarns are located on the top of a subsidiary volcanic plug in the Gorges Valley. Lake Michaelson lies almost 1,000 feet vertically below them. Unlike so many other lakes or tarns on the mountain, these are unusual in that the water has no outflow. Since this water is collected largely by direct and occult precipitation from the small low cliffs, which rise steeply 50 feet above the water level, the only means by which water leaves them is by direct evaporation, and in this dry atmosphere water loss by this means is considerable. The high and low water marks were surveyed in July 1959 and compared with Hunting Clan and R.A.F. aerial photographs; the maximum range was found to be four feet.

Due to the still nature of this water, *Subularia monticola* has colonised large areas of the bottom, while *Crassula granvikii* not only creeps over the sediment but in the very shallow water of Camp Tarn it forms a surface mat of vegetation. The wet banks of the Tarns support lush growths of the terrestrial forms of both these aquatic plants, together with mosses, liverworts and *Ranunculus oreophytus*.

Areas of damp boggy ground surround most of the Tarns, rising gently to steep cliff walls. The change of vegetation across this slope is very marked. The damp ground is covered with well developed tussocks of *Agrostis trachyphylla*, and where standing water is present *Carex monostachya* is dominant. Patches of bare ground support small clumps of the red-leaved *Luzula abyssinica, Alchemilla johnstonii, Geranium kilimandscharicum, Haplocarpha rueppellii, Haplosciadium abyssinicum, Ranunculus oreophytus* and *Swertia subnivalis*, some of which often creep over the tussocks.

As the damp ground gives way to a gentle slope, leading to shallow lichen encrusted cliffs, *Senecio keniodendron* and *Lobelia telekii* become common. The ground cover on small soil-filled ledges holds little bushes of *Helichrysum citrispinum* var. *armatum, Blaeria filago, Carduus platyphyllus* and *Cerastium afromontanum* ssp. *afromontanum*.

The rather rich ground flora at this altitude (14,000 feet) may be attributed to the absence of glaciers on this aspect of the peaks and hence the cold winds that occur in other valleys during the night are here virtually absent.

Tro Tarn (Peak Col. 14,500 feet)

This tarn is probably one of the oldest of the mountain's tarns, as it seems likely that this col was free of ice even at a time when ice was still well down the Teleki Valley.

Lying below Point Piggott, the Tarn and its surrounding vegetation are submitted to low night temperatures. In addition the shade afforded by the peaks means that in this situation sub-freezing temperatures may be expected for as much as sixteen hours per day. Water percolates through a damming moraine and flows down to the Teleki Valley where it meets the main Tyndal glacier stream. There appears to be a fluctuation of about 15″ in seasonal water levels, and below the lower level of this seasonal water level, no more water seems to flow out of the lake. The low temperatures and cold winds in this situation mean that all the plants occurring here are submitted to severe drying.

No *Subularia monticola* was found around this tarn, but *Crassula granvikii* grows around the edge of the ooze shelf although not so profusely as it has done near other tarns.

The bare open well-drained ground around the tarn supports low stunted tussocks of *Agrostis trachyphylla*, while sheltered areas

Plate 9. Glacio-fluvial flats at the head of the Teleki Valley: 14,000 feet. Note the moraine on the right, and the large areas of bare ground due largely to frost soil phenomena.

Plate 10. Specimens of *Ranunculus oreophytus* demonstrating the inhibitory effect of diurnal temperature changes in the Alpine Zone. The plant on the left has been maintained at room temperature in Nairobi, while that in the right has been maintained at room temperature during the day and frozen at night.

between rocks and close to the tussocks support small plants of *Arabis alpina, Carduus platyphyllus, Cerastium afromontanum, Nannoseris schimperi, Senecio keniophytum, Senecio purtschelleri, Valeriana kilimandscharica* ssp. *kilimandscharica.* On sloping moraine deposits above the tarn, stunted plants of *Senecio keniodendron, Lobelia telekii* and the coarse tussocks of *Festuca pilgeri* ssp. *pilgeri* enter the association.

Tyndall Tarn (14,600 feet)

The highest lake in this altitudinal series is this pro-glacial tarn at the foot of the Tyndall Glacier. Due to its close proximity to the retreating ice, the water is still milky from the large quantity of glacial flour suspended in it. The temperature of this water does not rise above 4°C, but it is still capable of supporting diatoms and crustaceans. The shore is lined with morainic rocks that give shelter to small groups of *Agrostis trachyphylla, Senecio keniophytum* and *Arabis alpina.*

5. The Nival Zone.

FRIES (1948) distinguished this as a separate zone, and based his separation on the disappearance of *Senecio keniodendron.* He placed the lower limit of the zone at approximately 14,929 feet (4,550 m). HEDBERG (1951) did not distinguish a separate zone at this level, but included it in the alpine zone.

The present author considers that the Nival Zone should be retained as a separate region on Mount Kenya. Strictly speaking, it is best considered as that area from which the glaciers have only recently retreated, and on which the earliest stages of vegetational colonisation are clearly discernable. Under the consideration of colonisation in this important region it will be pointed out that the Nival Zone starts abruptly at a point where signs of a recent glacial advance can readily be distinguished. This agrees closely with the Nival Zone as defined by TROLL (1958).

Since the glaciers have now retreated to an extent where they are mere ice fragments clinging to the face of the main peaks, the zone is fragmentary in arrangement and no longer the continuous belt that it must have been when a dome of ice covered the whole peak region.

The plants forming the main association of this region are small in number, and are all stunted and grow invariably in sheltered situations. The order in which the plants occur appears to be quite specific and will be dealt with under their colonisation.

In sheltered situations *Agrostis trachyphylla, Arabis alpina, Carduus platyphyllus, Lobelia telekii, Nannoseris schimperi, Senecio keniophytum* and *S. purtschelleri* occur here and there over the moraines, *S. keniophytum* being found within 25—30 feet of the retreating ice.

In addition to these regular constituents of the community, *Alchemilla argyrophylla, Helichrysum citrispinum* var. *armatum* and *Oreophyton falcatum* occur close to the point where the recent moraine deposits abut on to the older, lichen-encrusted material (i.e. Nival — Upper Alpine zone boundary).

Above the Nival zone *Helichrysum brownei* grows up to an altitude of over 16,000 feet. The author recorded these plants on Point Piggott at this altitude, and MACKINDER (1900) recorded the species on the main peaks. Although on an altitudinal basis this plant should be included as a constituent of the Nival flora, since they only occur in cracks on well-weathered rocky surfaces that have been free of ice action for a long period, it really occurs in what is an extension of the Alpine zone. Thus from an altitudinal point of view the Nival Zone may be considered as a distinct zone that occurs within the larger and broader Upper Alpine zone.

THE ALPINE CLIMATE

It has been suggested that the peculiarities of an Equatorial mountain climate (HAUMAN 1935, TROLL 1959) are in large part responsible for the unusual growth forms to be found in these regions. Until thorough climatological studies are made on these mountains, particularly with regard to such factors as ultra-violet and infra-red radiation, this question must remain largely one of conjecture. It is, however, possible to examine the climate of Mount Kenya, and from our somewhat sparse knowledge of this and other equatorial mountains, to try in part to correlate climate with the environment and the distribution of distinctive life forms.

Unlike mountains in higher latitudes, those on or near the Equator do not experience marked seasons, but they exhibit a remarkable diurnal fluctuation of temperature, which is a more or less constant feature throughout the year. Such a temperature regime, which HEDBERG (1957) so aptly calls "winter every night and summer every day", must have been a selective factor of great importance in the evolution of the Equatorial mountain flora. Certainly any plant which is capable of withstanding frost action on almost every night of the year, and intense radiation and insolation during the day, must have developed under conditions of rigid selection.

In such marginal alpine habitats it is not possible to separate climatic from edaphic factors, for the rate of development of the latter has largely been a function of the activity of the former. It will later be shown that the speed of soil formation on Mount Kenya is due chiefly to the action of frost and water. TROLL (1943, 1948) has called this remarkable alternation of frost and thaw, which results in so much frost lifting and solifluction, Frostwechselklima.

Our knowledge of the Afro-alpine climate is derived mainly from observations made by various expeditions that have visited these mountains. Such figures necessarily give a discontinuous picture of the climate in any one year. The author has been fortunate in obtaining access to a large amount of unpublished information, through the kindness of S. BRINKMAN, Meteorologist to the I.G.Y. Mount Kenya Expedition; R. JONES of the Hydrology Department of the Kenya Ministery of Works, and Dr. John GRIFFITHS.

Before considering aspects of the present climate on Mount Kenya reference must be made to the ancient climatic phases through which it has passed, for it is these past era's that have in large part been responsible for the evolution of the Alpine habitat that has been the subject of this study.

There has been a virtual revolution since 1960 in our understanding of climatic changes in the late Pleistocene. This knowledge has been derived from the pollen analytical studies of VAN ZINDEREN BAKKER (1962a, 1962b, 1964, 1965), LIVINGSTONE (1962) and COETZEE (1964). The close association of these studies with radio carbon dating techniques has enabled us to construct a time scale against which vegetation changes over the last 20—30 thousand years can be measured.

Pollen studies are particularly suitable for the Equatorial montane regions for their very distinct vegetation zones are very sensitive to small temperature and related humidity changes. In particular if the core for this type of study is removed from an area that is at the present time close to a vegetational boundary, its movement in relation to the fixed boring site is, to say the least, dramatic.

In 1962 VAN ZINDEREN BAKKER described a pollen profile from the Cherangani Hills, in N.W. Kenya. A core 3.7 m long was obtained from a swamp near Kaisungor (+1° N. lat., 35° 28′ E. long.) at an altitude of 2926 m (9600 ft). This site lies close to the edge of the montane forest, and above which Ericaceous moorland stretches to the highest point, 3445 m (11,300 ft). Thus at the present time these hills are not high enough to support the alpine grassland of other and higher East African mountains. Any grassland that is now present in this region is due to the influence of man and his grazing and burning activities. A sample from the lower end of the core (—2.85 —3.00 m) was dated and yielded an age of 12,650 ± 100 years BP.

Presuming an even rate of deposition the annual sedimentation increment of this swamp is .231 mm per annum. If the pollen profile is correlated with a time scale obtained in this way it has been possible for VAN ZINDEREN BAKKER (1962b) to demonstrate an interesting sequence of vegetation zone depression.

These changes pass from an alpine grassland vegetation 14,000 years ago, through forest margin indicator plants at 6000 years, which in turn grade into a forest maximum 3000—700 years ago. Since this time the profile suggests that a new cool period has again lowered the forest margin.

At the same time LIVINGSTONE (1962) reported on a core taken from Mahoma Lake (0° 21′ N. lat., 29° 38′ E. long) on the Ruwenzori Mountains at an altitude of 3000 m (9840 ft), which was taken at an altitude very similar to that in the Cherangani Hills by VAN ZINDEREN BAKKER (op. cit.).

The core, which was 6 m long, was removed from below 9.5 m of water and passed from organic sediment down to 5.4 m, through grey inorganic glacial sediment into coarse gravel. A sample removed from within .2 m of the bottom of the organic deposits was submitted to C14 analysis and gave an age of 14,700 ± 290 years BP. Not unreasonably LIVINGSTONE suggested that the presence at the bottom of this core of fine glacial debris suggested that 15,000 years ago ice was still in the vicinity of the lake's margin. This important discovery gave an approximate date for the deglaciation of this altitude on the Ruwenzori Mountains, which coincides well with the time at which it is known that the Wurm glacier of Europe and the Wisconsin glacier of America were undergoing a similar halting retreat.

More recently COETZEE (1964) has described a core from Mount Kenya. The boring was carried out in the Sacred Lake, which lies at an altitude of 2400 m (7572 ft), at 0° .03′ N. lat. and 37° .32′ E. long. on the mountain's N.W. flank. This site falls well within the montane rain forest zone and was therefore more ideally suited to detect vegetational depression than one of the higher alpine lakes.

The core studied was over 13 m in length, and a sample from the —12.30 m level was submitted to C14 age determination and yielded an age of 15,400 ± 180 years BP which is corrected by the author to 15,862 ± 185 years BP. The total age of the whole core was calculated as 18,626 years BP. The lower end of the core showed that at this time (18,626 BP) the lake was surrounded by Ericaceous scrub and alpine grasses. This vegetation shows a dramatic change at 10,000 years when there is a sharp decrease in grass pollen and a corresponding increase in *Hagenia*. Since this latter Rosaceous tree occurs in the upper limits of the forest it acts as a very efficient indicator of the approach of the tree line to the lakes margin. From this point to the top of the core there is a gradual increase in montane rain forest species and *Podocarpus*.

It should be remembered that this core was taken on the mountain's NE flank where the moorland is very extensive, and the montain rain forest although beginning to taper down from its maximum in the E-SE., is still very thick. The dramatic changes demonstrated in this core may well be due to the fact that this quadrant of the mountain is more sensitive to slight temperature and humidity changes.

The results of these three studies show the relation of the observed vegetation changes to both their age and the European climatic chronology. It should be noted that the Mount Kenya profile was derived from a site 600 m lower than that of Cherangani and Ruwenzori. If in relation to the other two profiles we correct the altitude of the Mount Kenya core it will be noted that a considerable degree of similarity is achieved.

The level of the dated section of the Sacred Lake shows that at this time the vegetation around the lake's margin was similar to that found at the present time 1000-1100 m (3280-3600 ft) higher. A depression of this magnitude would, if calculated on the same basis as did v.d. HAMMEN & GONZALES (1960) for the Equatorial Andes, have led to a drop in temperature of as much as 8° C. Whether lowering the temperature by this amount would have led to the lower edge of the forest being depressed by a similar amount is still a matter of conjecture. In any case even if the lower margin had only been lowered by 600-700 m (1970-2300 ft.) the montane and highland forest of the present time would have covered the whole Kenya Highland regions, thus rendering the now isolated forests of many of the smaller mountains in a state of connection. When CHAPIN (1934) suggested that the evergreen forests had in the past been 50% more extensive than they are at present, he may have been nearer the truth than he imagined.

In the light of these new discoveries it is of great interest to re-examine MOREAU (1952) and to note that many of the problems he posed have been answered, and that the answers obtained agree clearly with his own suggestions on the subject of Pleistocene climatic regimes.

1. Temperature.

The main figures for the Alpine Zone of Mount Kenya are those recorded as part of the I.G.Y. Mount Kenya Expedition's Meteorological programme, together with sporadic recordings made by the author and his colleagues on visits to various parts of the mountain. In addition, the East African Meteorological Department has, over the last few years, built up a picture of the variation of temperature with altitude. It will be seen that these figures, which are calculated on a mathematical basis, agree very closely with those actually recorded on the mountain. On a statistical basis it was found that, given the height of a location, its temperature might be calculated with a probable error of only 2°F* (Met. Dept. 1959, Part I; Kenya).

On examining the generalised Altitude/Temperature graph (Fig. 8) it will be seen that the fluctuation in mean temperature over the year at 10,000 and 17,000 feet is 3°F, while the annual range of maximum temperature at these altitudes is approximately 10°F. The original figures (Met. Dept. 1959, Part I; Kenya) show an increase in the range of maximum temperatures from 6°F at sea level to 10°—11°F at 17,000 feet. The mean annual minimum temperature figures show an amplitude of from 3°—5°F from sea level to 10,000 feet, with a slight rise to 7°F at 17,000 feet.

Temperatures recorded on the mountain by the I.G.Y. Expedition were all taken in the Teleki Valley. Recording Stations were distributed as follows:

---

* 1°F = $^5/_9$°C; temperature in °C = $^5/_9$ (°F–32).

(1) *10,000 feet* (3,048 m) Old Meteorological station on the Naro Moru track. An artificial clearing in mixed *Podocarpus – Arundinaria* forest (I.G.Y. Camp I).

(2) *13,750 feet* (4,191 m) MACKINDER's Camp at the head of the Teleki Valley. Short tussock grassland; station here slightly sheltered by an extensive terminal moraine (I.G.Y. Camp III).

(3) *15,650 feet* (4,770 m) Top Hut, situated on broken rocky ground, approximately 100 yards from the Lewis Glacier. Bare rocks and ice, devoid of vegetation. (I.G.Y. Camp IV).

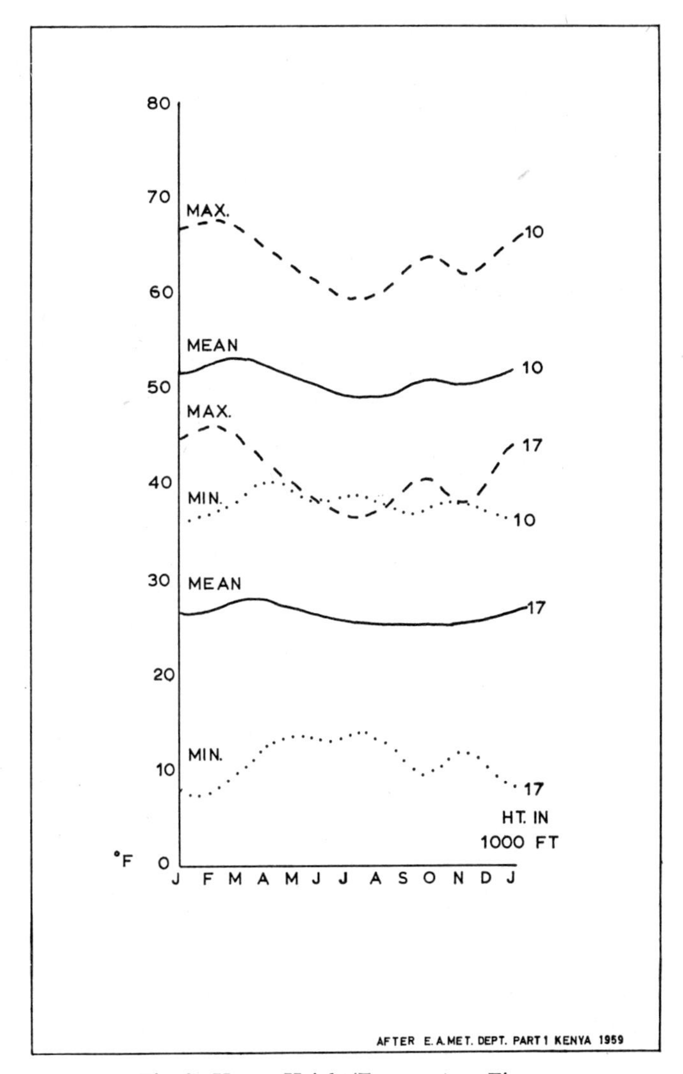

Fig. 8. Kenya Height/Temperature Figures.

Since it appears that the Alpine climate remains fairly constant throughout the year, the following figures obtained between December 19th, 1957 and January 17th, 1958 give a fairly clear picture of the temperature regime:

(1) *10,000 feet* (3,048 m)

Mean temperature	45.33 °F
Mean maximum	61,1 °F
Mean minimum	35.1 °F
Mean daily range	26.0 °F
Maximum daily range	37.0 °F
Minimum daily range	14.0 °F
Maximum temperature recorded	67.0 °F
Lowest temperature recorded	29.0 °F

(2) *13,750 feet* (4,191 m)

Mean temperature	35.51 °F
Mean maximum	41.6 °F
Mean minimum	25.5 °F
Mean daily range	16.1 °F
Maximum daily range	27.0 °F
Minimum daily range	13.0 °F
Maximum temperature recorded	51.8 °F
Minimum temperature recorded	20.0 °F

(3) *15,650 feet* (4,770 m)

Mean temperature	28.6 °F
Mean maximum	34.1 °F
Mean minimum	24.9 °F
Mean daily range	9.2 °F
Maximum daily range	18.0 °F
Minimum daily range	5.0 °F
Maximum temperature recorded	41.0 °F
Lowest temperature recorded	17.0 °F

Over the period for which temperatures were recorded at all three Camps simultaneously, the differences in mean temperature between Camp I and Camp III, and Camp III and Camp IV were respectively 9.85 °F and 6.89 °F, giving a lapse rate over this altitudinal range of 2.62 °F and 3.62 °F per 1,000 feet. The latter reading agrees closely with PEARSALL'S (1950) reference to 1 °F for every 300 feet. (The International Standard Atmosphere has a lapse rate of approximately 6.5 °C/km or 11.5 °F/km, being equivalent to 2.1 °C or 3.78 °F per 1,000 feet). The slight increase over the normal rate at the highest level is probably due to the cold glacier winds to which most areas over 15,000 feet are submitted. The low figure of 2.62 °F per thousand feet over the range 10,000—13,750 feet is a difficult one to explain, but it may be due to an inversion effect that seems to occur at low levels in Kenya during the dry season.

The figures cited above may be compared with the following table of upper air temperatures recorded above Nairobi and averaged for the years 1948—1955:

	03.00 GMT	06.00 GMT	12.00 GMT
Surface 5,360′	54.4 °F	63.8 °F	76.8 °F
800 mb. 6,710′	65.3 °F	56.7 °F	67.2 °F
700 mb. 10,380′	47,3 °F	47.7 °F	49.3 °F
600 mb. 14,520′	35.4 °F	35.5 °F	35.5 °F
500 mb. 19,280′	22.3 °F	21.7 °F	22.1 °F

(GMT = local time -3 hours)

For altitudes up to 10,380 feet it will be noted that the temperature rises steadily from 03.00—12.00 hours GMT, while over the same period, at 14,520 and 19,280 feet the temperature remains remarkably constant. Not surprisingly the lapse rate over what is more or less the equivalent of the alpine zone (10,380′ to 14,520′) falls steadily with time from 3.47 °F through 3.39 °F to 3.0 °F.

Mani (1962) gives some temperature measurements for Great Himalaya taken in June, at 4,200 m. It is interesting to compare these with the figures taken from a comparable altitude (4,191 m) on Mount Kenya, and with those recorded by Klute (1920) on the Machame Escarpment of Kilimanjaro between August 19th and October 12th, 1912 (Salt, 1951):

	Great Himalaya (4,200 m) 13,776 ft.	Mount Kenya (4,191 m) 13,716 ft.	Kilimanjaro (4,135 m) 13,563 ft.
Mean Maximum	12 °C	5.4 °C	5.17 °C
Mean Minimum	–1.5 °C	–3.6 °C	–0.77 °C
Mean Range	14 °C	8.9 °C	10 °C
Lowest	–10 °C	–6.7 °C	–4 °C

It will be noted that there is some measure of agreement between the three sets of figures. Those for the Himalaya show a higher mean maximum and also a higher mean minimum, although this range lies between latitudes 29° and 30° North. The greater range exhibited and the very much depressed lowest temperature are indicative of a greater seasonal fluctuation and much less distinct range. Also the high mean maximum figures recorded on Himalaya reflect again the importance of intense summer insolation in this region. The two equatorial mountains show a large degree of correlation; small differences may be reflected in the nature of the site at which the temperatures were recorded.

Although the mean maximum temperatures for the Alpine Zone

would appear to be low, these represent air temperatures above the ground, and do not take into account the unique microclimate close to the ground. GIEGER (1957) refers to the importance of this factor in quoting the work of MAURER, who found the amount by which ground temperature exceeds that of the air increases with altitude. Undoubtedly this, more perhaps than any other factor, is one of the prime controllers of alpine vegetation, for although the most obvious plants occurring in the Alpine Zone are those which exhibit megaphytism, by far the larger percentage of flowering plants are low, creeping or sessile forms, that live within the zone of influence of the ground climate.

Measurements made by SCAETTA on June 19th, 1929, North of Lake Kivu at Karasimbi, at 14,780 feet (4,506 m) are as follows:

12.00 hours	Free Air	5 °C
	Inside *Alchemilla* clump	14.6 °C
	Top layer of dry soil	16.2 °C
	Difference in ground and air	11.2 °C
13.00 hours	Free Air	3.5 °C
	Clumps of *Poa glacialis*	17.4 °C
	Dry lichens on a lava plateau	19.4 °C
	Difference in ground and air	15.9 °C

Measurements made by KLUTE (1920) on Kilimanjaro at 4,150 m illustrate a similar phenomenon. Here the difference between the air and surface temperature was as much as 31.20 °C. The actual range recorded largely depended on the density and nature of the ground cover. It is only, however, when this is considered as a diurnal phenomenon that its full significance is appreciated, for when a difference in temperature of up to 30 °C is recorded at noon, this will be reduced to almost nil at 19.00 hours. Thus, to obtain a true picture of the climate in the alpine zone, both air and surface temperatures should be recorded simultaneously.

In January 1963 air and ground temperatures were recorded at 13,700 feet in the Mackinder Valley. Figures were obtained by using thermistors, mounted at intervals on bamboo poles, simultaneous readings were taken from mercury thermometers, similarly mounted. These thermometers were insulated with cotton wool and their bulbs shaded by small curved plates of aluminium. The weather over the period was very cloudy, with intermittent rain, hail and snow. The results, however, although they do not show such a large air to ground difference as that recorded on Kilimanjaro, are nevertheless of great interest. Fig. 9 demonstrates clearly that one of the most important factors in the Alpine Zone's temperature regime is the great speed at which temperatures near the ground fluctuate. The site at which the recordings were made was shaded by a large ridge, and the sun did not reach the floor of the valley

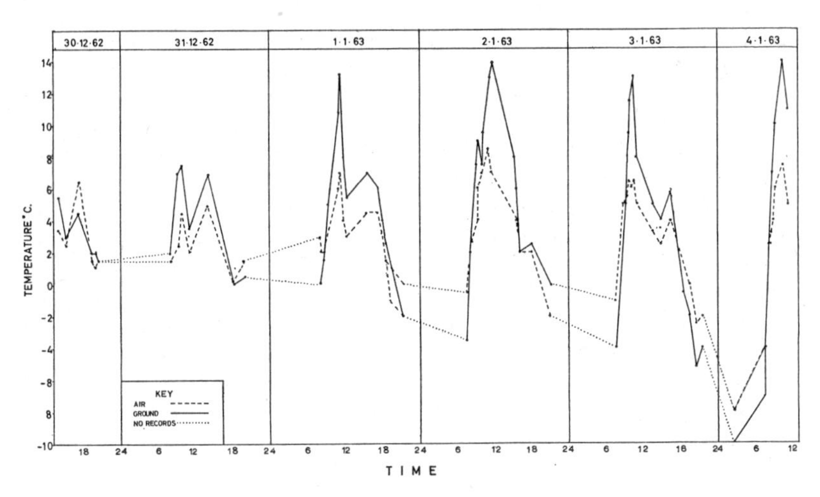

Fig. 9. Air to ground temperatures. Mackinder Valley 13,700 feet.

until 7.50 a.m. Provided that the sky was clear, the temperature rose rapidly, with the ground temperature rising sharply away from that of the air. When the sky became clouded, this fell rapidly, until the air and ground temperatures were almost equal. As an almost constant diurnal phenomenon, its effect on the ground vegetation must be considerable.

This remarkable contrast between the micro- and macroclimates of these areas is expressed even more clearly by MANI (1962), with measurements made on Great Himalaya (no altitude given).

Site	T °C	RH%
Atmosphere	-1.5 °	40%
5 cm above rock	26 °	12%
Rock surface	30 °	–
Under barren rock	10 °	98%
Fissure opening to air	18 °	70%

Although MANI's main concern was the effect of microclimate on insects, it is clear that it will also have a profound effect on the vegetation. In particular, it will tend to affect the growth form of plants to the extent that they remain within the sphere of influence of the climate near the ground.

The lowest temperature recorded in the lower alpine zone was 17°F (—8.3°C), noted by BAKER (personal communication) while camping on the Thego River at 13,000 feet in March 1959. On this occasion he noted that all the rosettes of *Senecio brassica* completely closed their leaves. This phenomenon has not been observed to this degree elsewhere on the mountain although low temperatures such as this are probably of frequent occurrence in frost hollows on other parts of the mountains. Unfortunately, over such difficult terrain, little is known of local climatic conditions. Heavy frost has been experienced by the author at 10,000 feet on the Naro Moru track (December 1956, January 1957), and in August 1961 on the Sirimon track at the same altitude, heavy frost was observed on five consecutive nights. SALT (1951) recorded his lowest temperature on Kilimanjaro, —2°C (28.4°F) at 12,450 feet, where he speaks of a heavy frost occurring every night.

In an environment where the rate of frost change exceeds 300 per annum (TROLL 1943), the effect on the life of these regions must be profound. Indeed, one may reasonably postulate that this alone must have been the most important single factor in the development of the unusual life forms to be found in the Equatorial mountains.

The lowest temperature recorded by the author was —9°C (15.8°F), at 13,700 feet in the Mackinder Valley in January 1963.

2. Rainfall.

Precipitation data for Mount Kenya has been more fully catalogued than any other climatological feature. This is chiefly due to the Hydrology Department of the Kenya Ministry of Works, who have over two periods erected rain gauges on the mountain. On the first occasion, gauges were placed along the Sirimon and Naro Moru tracks and records were taken for the period 1950—1952. Unfortunately, readings from this series of gauges were discontinued at the start of the Mau Mau emergency, as a number of terrorists took refuge on Mount Kenya. Later the Department kindly allowed Mr. R. JONES to accompany the I.G.Y. Mount Kenya Expedition as its Hydrologist in December 1957 and January 1958, and as a corollary to this work, a very complete series of gauges was installed giving the mountain a complete cover in almost all valleys, and at all altitudes. The 1960 Isohyets (Fig. 10) were prepared by Mr. JONES from data collected from these gauges.

The figures for the years 1950—1952 have been prepared as histograms, and are shown for the Naro Moru and Sirimon tracks at altitudes of 8,000 feet, 10,000 feet and 14,000 feet (Fig. 11). Examination of this diagram will illustrate the following points of interest:

(1) The main rainfall periods, although varying greatly in amount, do so in two fairly well defined seasons. The maxima occur in the months April—May, and November—December, and follow with reasonable accuracy the incidence of the lowland rains.

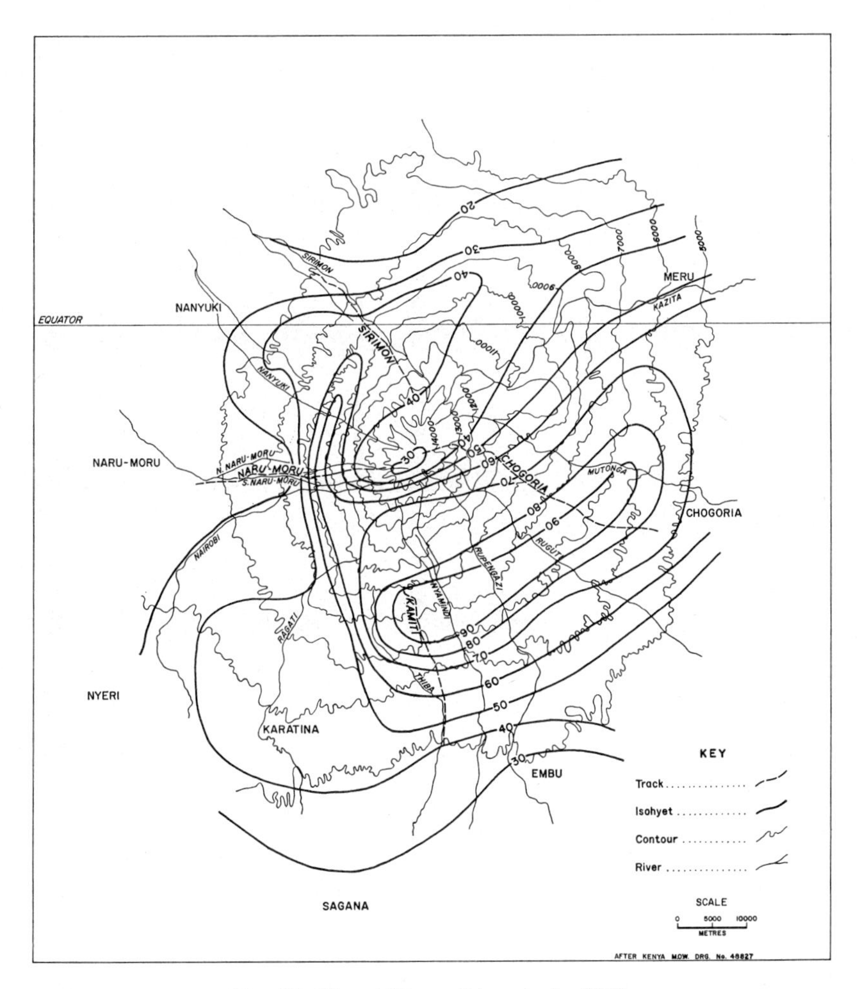

Fig. 10. Mount Kenya Ishoyets for 1960.

(2) Although the Sirimon track is in an area where the forest tends to thin out towards the Northern gap, there is correlation between its rainfall pattern and that of the Naro Moru track, where the forest is still quite thick.

(3) The annual variation in the "wet" periods appears to be considerable.

(4) Two distinct dry periods occur between the rainfall maxima, which seems to be most pronounced in the months January—February, and June to September. From the author's experience, although most plants have been recorded in flower throughout the year,

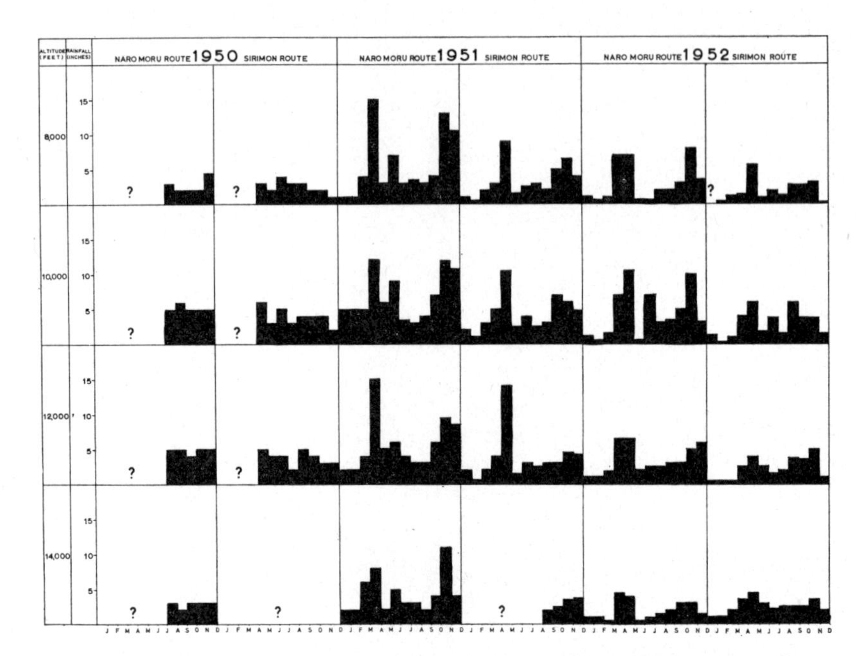

Fig. 11. Rainfall on the Western side of Mount Kenya.

mass flowering is most common in January and July (i.e. following the main rainfall maxima).

The Isohyets prepared for the year 1960 are the most valuable source of information on rainfall. It has been implied above that the amount of rain falling in the maximal precipitation periods varies considerably from year to year, but this ought not appreciably to affect the Isohyet pattern, merely their amplititude for any particular year. It will be seen that their maximum rainfall occurs in the South-East quadrant of the mountain, between 13,000 feet and 5,000 feet. This correlates well with the presence of the widest and most dense forest on this side of the mountain. It is not, indeed, surprising that early explorers who attempted to scale the mountain from this side never succeeded in traversing the forest. To the West the rainfall grades off in the montane forest zone to between 50″ and 60″ at due West, and to 30″ in the North. Again this explains the distribution of forest which occurs in an arc around the mountain, with its greatest width in the South-east and a gap in the North. These figures suggest that an annual rainfall of over 50″ is needed to support a pronounced montane forest belt, and that 30″ and below will produce the moorland scrub so characteristic of the Northern face.

If the Alpine Zone be considered as approximately the area over 12,000 feet, the rainfall here is only moderate. The main peak area has a rainfall of between 30″ and 40″ per annum, the 30″ Isohyet being displaced slightly to the South-west of the peaks. To the South and East the rainfall rises rapidly to between 60″ and 70″ per annum. It is on this aspect of the mountain that the Ericaceous moorland is particularly dense and where the limits of even the *Hagenia-Hypericum* Forest are considerably raised. The Gorges Valley is included in this area of higher rainfall which, together with the sheltering effect provided by its steep walls, accounts for the higher altitudinal rangs exhibited by *Erica* and other shrubs in the region.

SALT (1951) gives some interesting comparable rainfall data for Kilimanjaro, which are reproduced below:

Gauge	Altitude	1945	1946	1947	1948	1949	Average
1E	7,200 ft.	67.12	44.17	NR	70.58	68.86	62.68
2E	9,400 ft.	73.06	62.37	71.23	91.44	58.23	71.27
3E	12,500 ft.	37.60	19.76	21.95	21.47	18.83	23.92
4E	14,000 ft.	7.78	9.95	3.59	9.47	3.28	6.81
5E	16,000 ft.	1.75	4.19	1.47	4.60	2.80	2.96
6E	19,000 ft.	0.07	0.60	0.30	0.35	0.10	0.28
1W	9,950 ft.	NR	38.93	63.55	61.07	48.92	53.12
2W	13,150 ft.	NR	28.38	49.91	43.13	38.46	39.97
3W	15,350 ft.	NR	6.93	0.97	1.31	0.29	2.38

These figures compare fairly closely with Mount Kenya, with the annual precipitation rate rising sharply to a maximum at 8,000 feet and then falling off rapidly. The maximum on the South-west flank of Kilimanjaro is somewhat higher than elsewhere, since on this quarter the moisture-containing clouds ascend much higher before they meet the drying trade winds (SALT 1951).

The upper alpine zone of Kilimanjaro is, however, much more barren than comparable altitudes on Mount Kenya. This may be explained by the steep glaciated and sheltered valleys of Mount Kenya, which are absent on Kilimanjaro, and also by the pervious nature of the terrain of the latter. There are few places where free water can be found above the height of 12,000 feet, and hence the vegetation in these regions is nowhere dense.

Although the rainfall of the upper alpine zone is so low on Kilimanjaro, it is interesting to note that SALT (1951) considers that occult precipitation, although not measurable in terms of rainfall, is of great importance to both the flora and fauna. This is undoubtedly also true for the alpine zone of Mount Kenya, where the amount of dew on the vegetation every morning is considerable. The total amount of water derived either from dew or melting frost must be quite large. It is hoped that at some time in the near future it may be possible to devise some simple but accurate means of measuring this water on Mount Kenya.

The role played by humidity in the climate of the alpine zone is difficult to assess. The mean hourly humidity over the periods recorded on the mountain is fairly constant, with a slight rise in the late afternoon and evening. The humidity on any one day varied considerably with regard to cloud cover and wind. It was found that when a steady breeze blew and the sky was clear, the humidity was low, while in comparatively still conditions and with low cloud cover, there was a correspondingly sharp rise in humidity.

Recordings of relative humidity near the ground, made in January 1963, showed how rapidly the relative humidity changes. In dull and cloudy weather the figures taken at 6 cm and 26 cm above ground level, with an Edney Hair Hygrometer, ranged between 58% and 71%. When the weather was clear, however, the relative humidity fell to below 20% and remained at this level until the sky clouded over. The effect of this low humidity is only too evident on the faces of climbers which, within a few days of fine weather, become dried and cracked. When the sun sets below the horizon, the rapid lowering of temperature is accompanied by an equally sharp rise in relative humidity.

Cloud and mist comprise a feature of some importance in the alpine zone, though not apparently of such significance as on the mountains of the Northern Hemisphere and the Equatorial Mountains of the Congo. In July 1959, after a period of two days of almost continuous drizzle, the rain stopped and a heavy mist covered the area in which the author was working (Hall Tarns, Gorges Valley, 14,100 feet). This low cloud shrouded the area for three days and resulted in a low temperature, high humidity, and a heavy dew that persisted for the whole period of cloud cover. Again it may be said that such conditions must be important in providing surface water, but the relative effect of such continuous occult precipitation on the vegetation is difficult to assess. Work is at present being conducted in Kenya on ways and means of measuring this water, and it is hoped that these techniques, when developed, will be used on Mount Kenya.

The major importance of humidity must be the variation between that of the macroclimate and the microclimate associated with the vegetation. This great contrast in relative humidity levels is well illustrated by figures (Mani 1962) for Great Himalaya (page 59).

3. Wind.

Although typical up and down valley winds occur in a diurnal

Plate 11. Soil polygons in the Gorges Valley: 14,100 feet.
Note the orientated striping of the polygons which corresponds to the direction of the prevailing wind.

Plate 12. A severely frost heaved ridge top in the Kazita Valley which is almost completely devoid of vegetation.
The surface is so badly frost heaved that all vegetable growth is either reduced or completely suppressed.

cycle in the alpine zone, they do not appear to play a significant role in exerting any profound effect on the vegetation. This may be largely attributed to the dissected topography which affords a degree of protection to most plants. Nowhere is it possible to observe effects of wind such as can be seen on the saddle of Kilimanjaro (15,000 feet), where the scattered bushes of *Helichrysum newii* are completely devoid of flowers on the windward side, while they blossom profusely on the lee side of the same plants. Only in the region of the glaciers and ridge tops on Mount Kenya does the need for protection from wind become really apparent. In these situations *Senecio keniophyton, S. purtschelleri* and *Arabis alpina* are found exclusively on the lee side of boulders, where they are afforded a large degree of protection from night glacier winds.

4. Climate and the Alpine Vegetation.

TROLL (1960) has noted the remarkable similarity between soil conditions, plant forms and vegetation types to be found in the high mountains of the world and the sub-antarctic belt. The most obvious growth forms found on Mount Kenya are the striking megaphytic plants and the high percentage of the total alpine flora that exhibits a rosette growth form. The former show marked similarities with megaphytic types found on high mountains in other parts of the world. Indeed, the similarity between the growth forms of the Giant *Senecio keniodendron* and *Lobelia telekii* of Mount Kenya with the *Espeletias* and *Lupinus alopecuroides* is a remarkable example of parallel morphological development under similar climatic conditions. The rosette habit is more widely distributed and appears to be characteristic of both Arctic and Alpine environments.

If it is possible to find the common denominator of these habitats, to account for their production of similar growth forms, it must surely exist in their climatic conditions. TROLL (1958) has, in fact, clearly shown that the occurrence of similar vegetation belts on the world's high mountains may be almost wholly explained in terms of their climates.

It has already been shown that the most marked feature of the Afro-Alpine Equatorial climate is the wide range of diurnal temperature change. The peculiarities of this temperature change are reflected in the nature of frost action at the soil surface, where the frost/thaw cycle results in the formation of miniature soil polygon structure soils. Such drastic climatic factors make colonisation extremely difficult and, in addition, produce a strong selection pressure in favour of those plants that can adapt their growth forms to cope with these changes. On Mount Kenya these adaptations are largely expressed as megaphytic and rosette vegetation. The similarities of vegetation on high mountains of the world are further complicated when it is remembered that similar types of vegetation are to be found in subarctic regions, although here the temperature shows a small diurnal and annual range. The rate of frost change, however, is in the region of 236 days, and it seems that this is an important factor in producing similar types of vegetation to those found on or near the Equator. As TROLL (1960) has remarked in comparing the Southern cold temperature zone and the tropical high mountains, "All types of growth form in both zones suggest that the perpetually low temperatures have been a decisive ecological factor responsible for a series of adaptational characteristics".

The persistent occurrence of rosette or cushion forms in the arctic, sub-Antarctic and high mountains of the world may be explained by a very low growth rate, resulting in the production of extremely short internodes. It has been suggested by WALTERS (1956 p. 44) and others that the intense illumination in these regions is largely responsible, through inhibition, for the production of these shortened internodes. On Mount Kenya, although at ground surface the radiation during the day is high, the drastic temperature changes to which the vegetation is submitted are even more marked.

In January 1959, specimens of *Ranunculus oreophytus*, *Haplosciadium abyssinicum*, *Haplocarpha rueppellii* and *Crassula granvikii* were transported from an altitude of 13,700 feet to Nairobi (5,600 feet), where they were maintained in the laboratory at room temperature (± 24 °C). After ten days the leaves of all these plants had elongated to twice their normal length. The flowers which are normally completely sessile, also elongated and rose aloft on pedicels up to two inches long. After a further week, the short stem had begun to elongate and the internodes themselves began to grow. In correspondence with Dr. O. HEDBERG in Uppsala, Sweden, to whom the author had sent living material, he confirmed the same observations for the species enumerated above, and added *Conyza subscaposa* and *Nannoseris schimperi* as also exhibiting the same exaggerated growth at room temperature. In January 1963, specimens of *Ranunculus oreophytus* and *Haplocarpha rueppellii* were again brought to Nairobi, where each clump was divided into two equal parts, one being maintained at room temperature while the other was kept at room temperature during the day and at night was transferred to a refrigerator, thus providing approximately the same conditions of marked diurnal temperature changes to which these plants are submitted in the alpine zone. Each day the leaves and, in particular, the pedicels of new flowers were measured. It was found that the pedicels of plants maintained at room temperature elongated rapidly, while those that were cooled at night showed a barely perceptible increase in length. After 14 days the plants were photographed (see Plate 10) and it can readily be seen that the plants under a controlled temperature (i.e. those on the left) are nearly three times as long as the specimens (on the right) subjected to the diurnal change. This simple experiment does seem to indicate that temperatures are the prime controlling factors of this particular growth form. It is also of interest to note that although these plants must have grown in the alpine zone for a very long period, they are able to revert to an erect habit within a few days of removal from one of their main controlling factors.

The high percentage of rosette forms occurring in the Arctic, where long periods of daylight occur during the "midnight sun", had been quoted as an example of intense illumination causing inhibitation of internodes. While this is undoubtedly a factor of great significance in this environment, the temperature regime must also play a very important part. At Kiruna, in Northern Sweden, the period of vegetable growth is about 3½ months per annum, from the beginning of May to the middle of September, and the mean monthly temperature rises above freezing point for about four months only. In addition, snow may cover the ground for about eight months of the year, while the hours of sunlight vary from 300 hours in June to between six and ten hours in January. Hence, although annual plants may exhibit fairly rapid growth during the midnight sun, non-woody perennials with a rosette habit stand a greater chance of survival during the eight months period of snow cover than plants with an erect form of growth.

While it seems, however, that the suppression of internodes is to a great extent a response to changes of temperature in the alpine zone, the effect of light is also of some, if not of considerable importance. Certainly the production of a large number of species with deeply pigmented, reduced, hairy or even coriaceous leaves is to a greater or lesser degree a response to light, and in particular to ultra-violet radiation. It has been shown in recent experiments (v.d. VEEN & MEYER 1959) that if a plant is irradiated with ultra-violet light, the leaves die and fall off, but that subsequent leaves produced by the plant, if the illumination is continued, are harder, smaller and more hairy, and often have a thicker cuticle. It has even been suggested that high altitude plants show more profuse growth when transported to lower levels, owing to the influence of a high intensity of ultra-violet radiation on the efficiency of photosynthesis in the plant.

In addition to the large number of low plants occurring in the alpine zone, there are the megaphytic plants, which are in many ways more difficult to explain. Perhaps the most striking feature of the Giant Senecios and Lobelias of Mount Kenya is the oligoblastic (few branched) growth of the former and the apparent monoblastic nature of the latter. The main feature of importance in the development of this growth would seem to be found by comparison with their counterparts in the montane forest, some 3,000 feet below. This is not to say that the alpine species cited are necessarily to be considered as deriving directly from montane forest species existing today.

The montane forest of Mount Kenya supports *Lobelia gibberoa* and *bambuseta*. Both these plants are very tall (up to 25 feet), slender and many branched. COTTON (1944) has suggested that the conversion of a tall, many-branched plant of this sort to an oligoblastic form may be by suppression of the growing points. This would seem to be illustrated by the megaphytic species of *Senecio* and *Lobelia* on Mount Kenya. A slender etiolated plant, although successful when protected by the canopy of the forest, would not survive long when submitted to the rigours of an alpine climate. The forest plants are less woody, and the leaves when they die do not persist as a thick frill around their stems as they do in the erect alpine species. The much larger and tighter rosettes of the alpine species is also undoubtedly another factor of great importance in protecting their growing points from frost damage.

The sessile alpine *Senecio brassica* has a similar habit to that of *Lobelia keniensis* and *telekii*, since all three occur as sessile rosettes from which erect inflorescences arise. It has already been pointed out that *Lobelia keniensis* occurs in "family groups". Examination of the closely associated rosettes of this species has shown that they arise from a robust shallow stem that closely resembles in detail the erect stems of the forest species *Lobelia gibberoa*. The dead rosette leaves of *Lobelia keniensis* are persistent and also serve to protect the stem from frost damage.

It would therefore seem that the megaphytic *Lobelia* and *Senecio* species of the alpine zone owe their success to this distinct modification: first by the suppression of growing points and the production of erect woody plants, protected by a persistent leaf frill, and secondly by the reduction of erect stems to shallow rhizomes from which only the much-thickened and pigmented leaf rosettes rise above the ground.

THE DEVELOPMENT AND DISTRIBUTION OF ALPINE SOILS

Mount Kenya's glacial history has already been mentioned and it must be referred to here as the major soil forming factor on Mount Kenya. It is true that the diurnal temperature fluctuations and other alpine climatic factors have played an important subsequent role in both particle formation and their distribution, but the primary agent of soil formation in all cases has been that of glacial action.

The Alpine zone is entirely volcanic in origin and the basic material from which soils have been derived are moraine deposits scattered in the wake of the retreating ice sheet. After the period of glacial maximum, when the ice dome that covered the upper part of the mountain began to retreat, ridge-top moraines, very strong lateral and terminal moraines, and finer ground moraine in the lower valley sides and floors were left behind. The material deposited at any one time from which soils could be derived depended on the stage and type of the glacier's retreat at that period. Thus, when the glacier retreated at a fairly even rate, boulders were scattered along the deeply scoured valley walls, while finer material was left in the valley bottoms. Due largely to their colloidal properties these finer particles were washed out rather slowly, but when they were so removed they settled out quite rapidly in the valley bottoms to form the very common clay soils of the region. Between these periods of retreat the glaciers remained remarkably steady and some extensive terminal and lateral moraines were laid down.

1. Soil Generation.

Just as the primary processes of flowering plant colonisation can be seen close to the glacial remnants of the present day, so the processes of soil formation are also very evident. The Tyndall glacier produces an ideal site for such observation, with its large terminal moraine and proglacial tarn. At the glacier's snout boulders of various sizes are continuously melting out of the surface of the ice, shedding an even layer of large boulders and gravel behind the moraine. Extensive glacial striae on all boulders in the area demonstrate the severe scouring action of the ice.

The rocks on the floor of a large ice cave at the foot of this glacier are thickly coated with glacial flour, and the ice scouring the surface of rocks still buried below filled the air with a powerful "flinty" odour. Evidence of the large quantity of these fine clay particles which are subsequently washed from these boulders is clearly shown by the milky colour of the Tyndall Tarn, where the water is always disturbed by a small glacial stream, and hence a large number of these particles remains in suspension. "Roche moutonée" forms a marked feature of this area. Due, however, to the relatively steep slope, the absence of stabilising vegetation and the erosion due to seasonal melting of snow, all the finer fractions that are carried out by water are deposited much lower down the mountain and do not form the glaci-fluvial fans that can be seen at Mackinder's Camp and below the Lewis glacier, where the terrain is much less permeable.

These newly produced glacial parent materials must be very similar to those from which all other soils in the alpine zone have been derived. The mechanism of soil generation in the region of the peaks is almost entirely by frost action, with water being the active transport agent once it is formed. The diurnal changes of temperature experienced by these rocks is considerable, and results in the continuous and complete shattering of even very large boulders. Flaking, however, from the surface of boulders forms the commonest frosting phenomenon, which is a prerequisite to colonisation of rocks in these regions by mosses and lichens.

On account of the low temperature prevailing at high altitude and the consequent paucity of soil organisms, there is a marked inhibition of the chemical breakdown of parent materials. This is well illustrated by the relative acidity of the alpine soils (pH 5.0—6.0), the low conductivity of aqueous soil extracts (45 mho) and the comparative purity of the stream and lake waters as demonstrated by Heinz LOEFFLER in 1961 (personal communication, 1962).

The presence of a high percentage of glacially derived clay particles gives the alpine soils a capacity to hold large quantities of water, resulting on flat or gently sloping ground in large areas of semi-bog. Such waterlogging is common in most valleys and is of great importance in the reduction of frosted rubble to fine particles.

Once solid particles have been produced by glacial and frost action, a series of new factors becomes important in the distribution of these materials. Time, together with the mountain's deeply dissected topography, is of great significance in distributing these weathered soil particles. As SCOTT (1962) has noted, in such well dissected areas the rate of transport of soil particles is accelerated and hence the soil will tend to remain youthful, due to the speed at which newly formed materials are removed. On the steeper valley walls such as the Teleki and Mackinder Valleys, there is a distinct graduation of gravity sorted material, while in the shallow regions such as the Nairobi Valley such sorting is negligible.

2. The Differentiation of Alpine Soil Habitats.

Soil movement and deformation are of great significance in governing the distribution of soil types and in their effect on vegetation. The principal inter-related processes observed in the Alpine zone are:

(1) Soil stone polygon formation:
(2) Soil heaved by needle ice:
(3) Solifluction:
(4) Fluviation:
(5) Stone "glacier" formation.

(*i*) *Ridge top soils.*

The Equatorial mountains are interesting for the frequency of frost soil phenomena. Such frost soils are generally associated with regions of permanent frost (GODWIN 1956), but in the Alpine zone of Mount Kenya they are the result of diurnal freeze and thaw changes, where, although the frost does not penetrate more than a few inches into the soil, its influence is sufficient to produce very typical frost structure soils. The frequency of soil polygons and stone stripes on Mount Kenya has been noted by TROLL (1949) and by ZEUNER (1945), being most common in valley heads between 14,500 and 15,500 feet, and on ridge tops where they develop to such an extent that all plant growth is suppressed.

These soil polygons are between 3″ and 6″ in diameter and are separated by stripes of stony material up to 1½″ across (Plate 11). Soil particles in the centre of the polygons are small and very much compacted, while soil below the plates (from 1″—1½″) is also compact, but appears softer and wetter than that in the plates. The shape of the soil polygons varies with the surface on which they are found, and on ridge tops and similarly exposed situations they are more regular in shape although their surfaces are frequently striped by wind. This is clearly illustrated in Plate 11, which was taken on slightly sloping ground in the Gorges Valley, at 14,000 feet.

On ridge tops, which are usually scattered with boulders and small stones, soil polygons are particularly common. In the more exposed situations, such as the broad ridge above Shipton's Cave in the Mackinder Valley (14,500 ft) these drastic sorting agencies make colonisation by plants almost impossible (Plates 17 and 18). The only plants found in these situations are lichens, which are thickly encrusted upon the larger boulders. Where even the slightest shade is afforded to patches of heaved soil, they are stabilised from their outer edges by mosses and late flowering plants (Fig. 12). At lower altitudes the ridge tops are well covered with grass tussocks and structure soils are not observed.

The actual casual agents of these polygons are difficult to discover, but the basic requirements (Troll 1958) for the formation of structure soils, which are adequately fulfilled on Mount Kenya, are intensive radiation and the very marked daily evaporation that accompanies it. No doubt the colloidal nature of the soil particles goes a long way towards the maintenance of water in the soil,

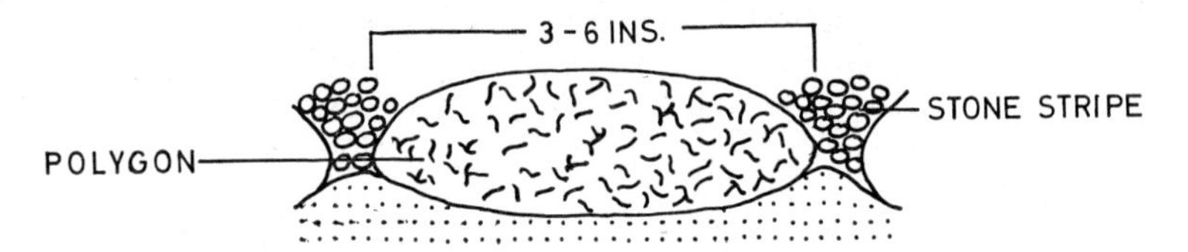

1. FRESH SOIL POLYGON

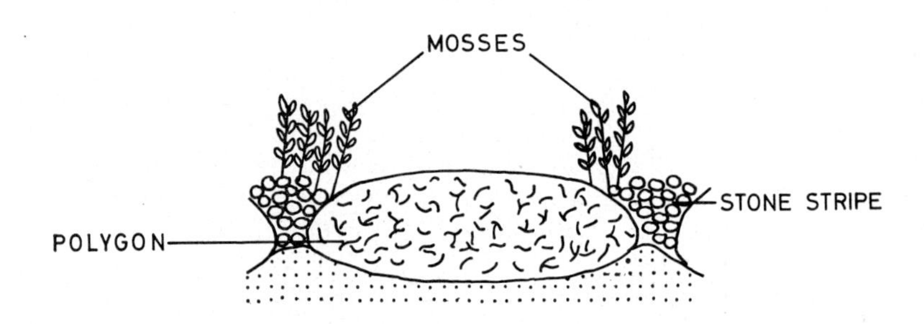

2. EARLY COLONISATION

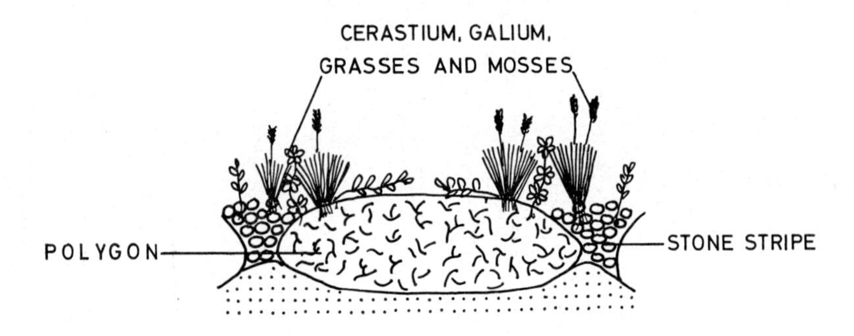

3. LATER POLYGON STABILISED

Fig. 12. Stabilisation of Soil Polygons

Plate 13. Severely frost heaved surface at 14,500 feet in the Mackinder Valley.

Plate 14. Mobile scree strip at the foot of the wall of the Mackinder Valley: 13,800 feet.
Passing through *Senecio-Alchemilla* forest.

Plate 15. Surface soil in the floor of the Mackinder Valley, badly disturbed by needle ice: 13,800 feet.

Plate 16. The same patch of soil shown on Plate 15, with some of the heaved soil removed to expose the ice crystals holding soil crumbs some 2½ inches above the more compact sub layer.

and in conjunction with this the evaporation from, and capillary attraction amongst the soil particles, causes the soil to remain continuously damp even over long periods when direct precipitation does not occur.

In areas where due to the steepness of the slope very little fine material exists, instead of the more characteristic polygons, small lines of crumbly soil, reminiscent of worm casts, are common. Such surfaces are extremely unstable and do not often become vegetated.

(*ii*) *Valley Side Soils.*

Valley sides leading from flat or knife-edged ridge tops face other important soil movement problems. The degree of movement in these situations is dependent upon the angle of slope and the degree to which the walls are broken by large boulders or lava outcrops. Again diurnal frost changes are the dominant climatic agency and here they lead to extensive soil heaving or solifluction. On valley walls throughout the alpine zone such moving soil surfaces are very common, and in places these areas are so large that they are completely devoid of vegetational cover. Where the surface is broken by lava outcrops, deep terraces of waterlogged soil have been produced and result in large scale vegetated stands on which may be found plant associations typical of this type of material in other wet areas. Where, however, the valley wall is not so broken, the surface is extremely unstable and solifluction carries material to the valley bottom where it overlays other older glacial deposits. On most valley walls one is especially struck by the marked inter-digitation of *Senecio keniodendron* which extends downwards, and by *Senecio brassica* which occurs at the foot of the slope and extends upwards in the form of thin and narrowing tongues. When these areas are examined, it is found that the latter are occurring on deep solifluction terraces that have accumulated at the foot of the wall, and that their progression upwards follows this terrace up the valley side. Where this vegetation stops, a thin band of scree usually continues towards the ridge top, representing in large part the coarser material of the valley wall from which the terrace has been washed out (Plate 14).

It is in areas where the passage of creeping soil is impeded that a good supply of surface water exists, and *Senecio keniodendron*, underlaid by dense stands of *Alchemilla argyrophylla* has become established. Once these areas are stabilised, streams meandering down the valley wall remove the finer soil particles and create narrow strips of very mobile scree. It seems probable that in many areas this type of secondary dissection of the valley walls is responsible, together with frost changes, for the production of the interesting pebble glaciers that are found in most valleys at about 14,000 feet.

The manner in which *Senecio-Alchemilla* forest thins out towards the ridge tops, and the presence of both dead *Senecio* and dead *Alchemilla* in these situations, suggests that the rate of soil removal is in excess of its rate of formation, and that this vegetational association is slowly being depressed down the valley wall.

Such depression is particularly obvious on smaller ridges as, for instance, on the terminal moraine found below Lake Höhnel. These deposits bear along their sides considerable areas of dead *Alchemilla argyrophylla* and *Helichrysum citrispinum* var. *armatum*, and such remains provide a good indication that they are slowly being denuded of vegetation owing to the removal of their fine soil fractions. At lower altitudes moraines and other ridges are less steep and are not affected by frost to the same degree as those at higher levels. In these situations the main transport agency is water, and due to the more gradual slope the finer fractions that have been washed out of the moraine ridges have formed platforms of heavy clay soil between their concentric walls. Small streams that cut through these deposits must account for a certain degree of leaching, but the high percentage of clay content of the soil no doubt cuts down these losses to a minimum.

(*iii*) *Valley Floor Soils.*

On the flat or slightly sloping ground of valley bottoms, an important feature of soil movement is that of Needle Ice, or Piprake. The formation of such crystals is associated with short periods of daily frost. On Mount Kenya these formations are particularly common on patches of bare soil in tussock grassland, at over 12,000 feet, and on larger patches of open ground such as the dried lake beds that are such a common feature in many valley heads, whose surfaces are disturbed every night and are completely devoid of vegetation. These crystals are formed at the interface of a drier surface layer and a damp sub-layer. This lower layer is usually well compacted and water is drawn up through pores in its surface to form plate-like ice crystals. These crystals lift soil crumbs up to $2\frac{1}{2}''$ above the compacted sub-layer (Fig. 13). Photographs taken at 13,700 feet in the Mackinder Valley (Plates 15 and 16) show the nature of these soil disturbances, and in oblique view, the ice crystals holding raised soil crumbs.

Grasses, mosses and other small plants are often heaved and broken in areas where needle ice is common, and although above an altitude of 12,000 feet needle ice occurs almost every night, the degree and severity of such lifting seems to vary. Such variation in the degree of needle ice formations appears to be related to the diurnal temperature fluctuation. This phenomenon was noted in the Mackinder Valley during January 1963, when it was observed that days of fine weather (i.e. high radiation and evaporation) were often followed by particularly low night temperatures, and on such occasions patches of soil that appeared to have been stabilised by vegetation were completely broken and destroyed by needle ice heaving. On the same occasion it was noted also that severe frost was more common on water washed terraces than on ground raised only two or three feet above this level.

Undoubtedly such frost heaving plays an important part in keeping the vegetation in valley bottoms in an immature state. On such ground the success of plants such as *Haplocarpha rueppellii, Haplosciadium abyssinicum* and *Ranunculus oreophytus* may be attributed to their stout tap roots, which make them less susceptable to such disturbances than the more delicate *Cerastium, Swertia,* etc.

In valley bottoms the action of glaciers and water as transporting and soil depositing agents can be seen. At Mackinder's Camp in the Teleki Valley the retreating ice sheet laid an extensive terminal

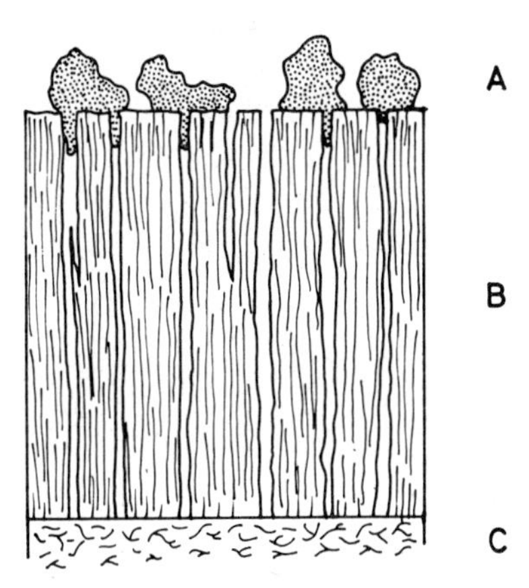

A Heaved Soil Crumbs
B Needle Ice Crystals in Plate like Bundles. Free of Soil
C Unfrozen, Compact, Structureless soil

Fig. 13. Frost heaving by needle ice.

moraine across the valley floor which appears to have been subsequently broken through. Behind this moraine, at this stage of the glacier's retreat, there existed either a lake or at least an area of swampy ground. At the present time only the glacio-pluvial deposits of this period are still evident. Further up the valley as the glacier recommenced its line of retreat a boulder-strewn ground moraine was laid down that still spans the valley floor. Such deposits have also been observed in the heads of the Mackinder and Gorges Valleys.

The stream below Mackinder's Camp in the Teleki Valley cuts through these glacio-fluvial deposits and demonstrates a thin layer of top soil overlaying extensive layers of yellow clay below. The clays show a certain number of darker stratifications but, due to the low temperatures, there are no signs of laterisation. (Fig. 14). It has already been pointed out that such deposits carry a characteristic vegetation which includes *Myosotis keniensis, Nannoseris schimperi* and *Cerastium afromontanum.*

Wind stripping of frost structure soils is not an uncommon feature in the Alpine zone. Soil polygons very frequently show fine orientated stripping. TROLL (1949) observed that needle ice occurring in the Gorges Valley was orientated in a South-east to North-west direction. This follows the direction of the prevailing

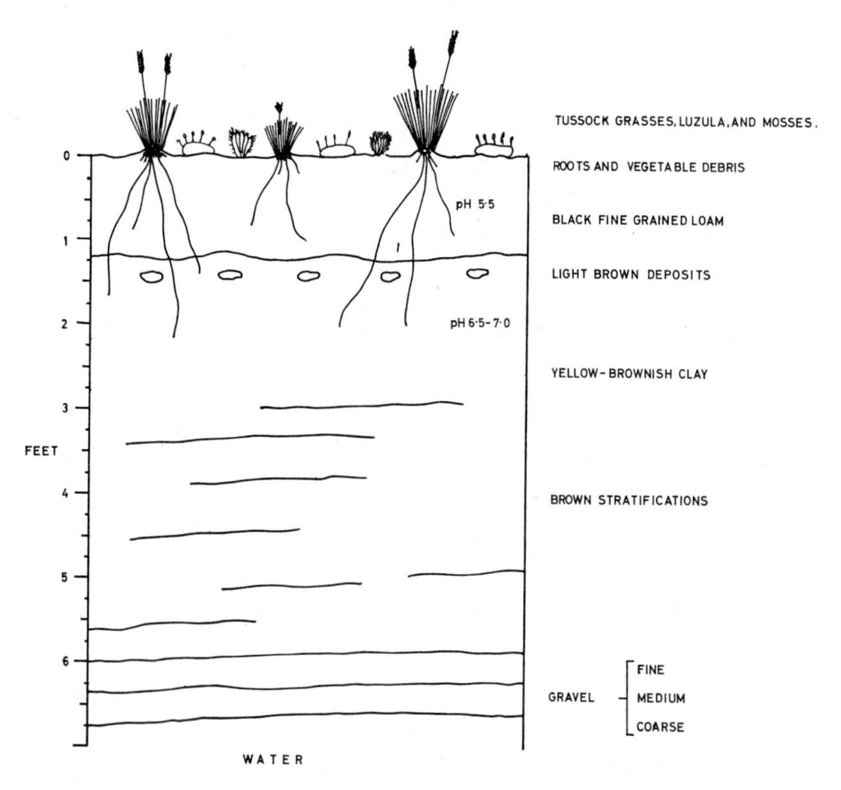

Fig. 14. Soil Profile. Glacio-Fluvial Deposit.
Teleki Valley 13,700 Ft.

winds which blow up the valley in the late afternoon, carrying cloud to the peaks. Thus in the valley bottoms particles derived from glacial action have collected and later, with frost heaving, the upper layers of these fine deposits become dried during the day and are subsequently carried by wind to form the eolian deposits of the peak region. ZEUNER (1945) studied the relations of frost and wind soils in the peak area, and found that their dominant fractions were .2 mm—.01 mm for eolian deposits and .05 mm—.01 mm for loess. The importance or extent of such deposits does not appear to be such a major issue on the isolated peak of Mount Kenya as on the more extensive Himalayan range. This fact is illustrated not only by the lack of accumulations of wind blown soil, but also by the paucity of wind blown insects that are so commonly associated with eolian deposits on other high mountains (MANI 1962).

3. The Structure and Chemistry of Alpine Soils.

Little information is available relating either to the structure or the chemistry of these soils. Through the kind assistance of Dr. BELLIS, of the Scott Agricultural Laboratories, Nairobi, a number of analyses have been carried out on soils collected by the author and others.

It has already been shown that soils associated with ridge tops have been subjected to accelerated erosion processes, and have in consequence lost their original characteristics. Soils that have been carried from ridges and valley sides may be considered under the general heading of Alpine meadow soil (SCOTT 1962). Examination of these deposits shows them to be very dark brown (10 YR 3/2), highly humic, with a carbon content of between 7% and 10%. The A horizon overlies a dark greyish brown (10 YR 4/2) to very dark brown, structureless, acid (pH 5—5.5) peaty loam, with scattered rocks. This type of soil is common from the forest to the peak region (9,000 feet to 14,000 feet), but it is thicker and better developed at lower levels, where the climate is less severe and more conducive to chemical breakdown.

The upper layer of these soils has a high humus content, but the rate of its formation must be greatly retarded at 12,000 feet. The slow rate of vegetable breakdown is clearly seen in the leaf frills surrounding the stems of the megaphytic *Senecio keniodendron*, which although the lower leaves are most probably up to 50 years old, show little sign of decomposition. Soils in the upper Mackinder Valley showed a carbon content in the top humus layer (i.e. 0—6″) of 12.59%. It will need a great deal more analytical work by a soil specialist before a decision can be reached, but it appears that top soils close to stream courses have a higher carbon content than those derived largely from solifluction at the base of valley walls.

Further analysis of samples collected at 13,700 feet revealed the following composition:

Mineral	13,700 feet Mackinder Valley: (base of valley wall)	13,700 feet Mackinder Valley: (close to stream course)
pH	5.2	5.2
Na m.e %	0.2	0.3
K.m.e. %	0.3	0.4
Ca. m.e. %	0.4	1.6
Mg. m.e. %	0.6	1.8
Mn. m.e. %	0.3	0.3
P.p.p.m.	52	58
N %	0.73	1.10
C %	8.76	12.59

The dionic conductivity of Alpine soils was measured on 1:5 aqueous suspensions for samples collected at 11,500 ft. and 13,500 feet in the Teleki Valley. These examinations gave the following results:

Alpine-Meadow Soil 11,500 feet.

Depths	0.4″	4″–12″	12″–22″	22″–28″
Dionic Conductivity	80	70	37	31

Glacio-Fluvial Deposits 13,500 feet.

Depths	0.6″	24″–36″
Dionic Conductivity	45	29

Without extensive studies on soil composition in the Alpine zone of Mount Kenya and other Equatorial mountains, it is difficult to suggest the possible effect of these low values on the vegetation. It is perhaps of interest, however, to note that similar figures were obtained by PEARSALL (1956) for the marginal impoverished soils of the Serengeti Plains, Tanganyika (i.e. 45 mho).

It is probable that low levels of exchangeable material in the Alpine soils may perhaps be best explained in terms of the low temperatures prevailing, the comparative insolubility of the lavas, and the speed at which materials once formed, are removed.

In addition the acid conditions prevailing throughout the zone result in a very small, and almost unknown soil flora and fauna. The following range of pH values were obtained over a wide altitude range in January 1958:

Altitude	Locality	pH
10,000 feet	Naru Moru Track	5.5
11,000 feet	Naru Moru Track Vertical bog	5.0
12,700 feet	Naru Moru Track Damp ground	5.0
13,700 feet	Mackinder's Camp: top soil	5.5
13,700 feet	Mackinder's Camp: sub soil (2′)	6.5
13,900 feet	Teleki Valley wall	5.0
14,500 feet	Teleki Valley head, frost heaved	5.0

In discussing vegetation communities it has already been pointed out that drainage plays an important role in their distribution. Ridge tops and valley walls contain a high percentage of material of large particle size, while glacio-fluvial deposits are composed of a high percentage of fine glacial flour fractions. The commoner alpine meadow soils that cover most valley bottoms, and other more or less well vegetated ground, were examined and found to contain a high percentage of fine sand and silt:

Alpine Meadow Soil: 11,300 feet.

Depth	0-4″	4″–12″	12″–22″	22″–28″
Coarse Sand %	4	2	25	9
Fine Sand %	48	52	53	43
Silt %	40	36	14	38
Clay %	8	10	8	10

COLONISATION IN THE ALPINE ZONE

After the height of the glacial maximum when the ice sheet began to retreat across the face of Mount Kenya, the primary stages of vegetational colonisation that later led to the establishment of the present plant associations must have taken place in its wake. Since the mountain still bears the remnants of these glaciers in the peak region, it would seem reasonable to assume that the stages of colonisation which may be observed in their vicinity at the present time must be very similar to these that took place when plants first became established in the Alpine Zone.

Although these early stages of colonisation may be observed in the Nival Zone, the intermediate stages that led to the well defined associations of the present time are more difficult to identify. This is largely due to the fact that the Upper Alpine Nival boundary occurs on steep, boulder strewn terrain, of considerably greater age than the more recent Nival moraines. The steep slope of this region and the great speed with which the fine fractions are removed constitutes another important factor that makes the picture of intermediate colonisation somewhat obscure. There are, however, several areas in which it is possible to see different phases of colonisation, and in almost all cases these phases are connected with the predominant soil distributing factors and the varied surfaces that these agencies create.

1. Primary Colonisation in the Nival Zone.

This zone is fragmentary in nature, being found not in a belt, as it must have been in the past when an ice cap completely covered the mountain, but as a series of crescentshaped areas at the foot of the present glaciers. As far as colonisation is concerned, only the Tyndal and Lewis glaciers proved suitable for vegetation studies. This is due to the fact that only these two ice masses are retreating on a fairly even front, while most other glaciers have now been reduced to mere stumps, clinging to the vertical cliffs of the peaks, so that their moraines slope steeply and their surfaces are in consequence extremely mobile, producing a surface which is quite unsuitable for colonisation. The position of glaciers and recent moraines in the peak area is shown in Fig. 5.

Field work in this region has produced some interesting data on colonisation, and at the same time has supported geo-glaciological opinion that a short period of advance took place little more than 100 years ago. This study was carried out as part of the International Geophysical Year Mount Kenya Expedition programme. In particular, the survey was conducted in collaboration with B. H. Baker, Geologist to the expedition (Coe 1959).

Plate 17. *Senecio keniophytum* growing on newly exposed moraine at the foot of the Tyndall glacier: 14,800 feet

Plate 18. Pro glacial Tarn at the foot of the Tyndall Glacier, held by a large terminal moraine: 14,600 feet.
Note the milky appearance of the water due to large quantities of suspended glacial flour.

The glaciers at present clothing the peaks of Mount Kenya provide an interesting region in which the order of plant succession on newly exposed moraine can be studied. These moraines occur at altitudes of between 14,422 feet (4,400 m) and 14,760 feet (4,500 m). It has already been stated that, due to the direct line of retreat of the Tyndall and Lewis Glaciers, and also because the rate of recent

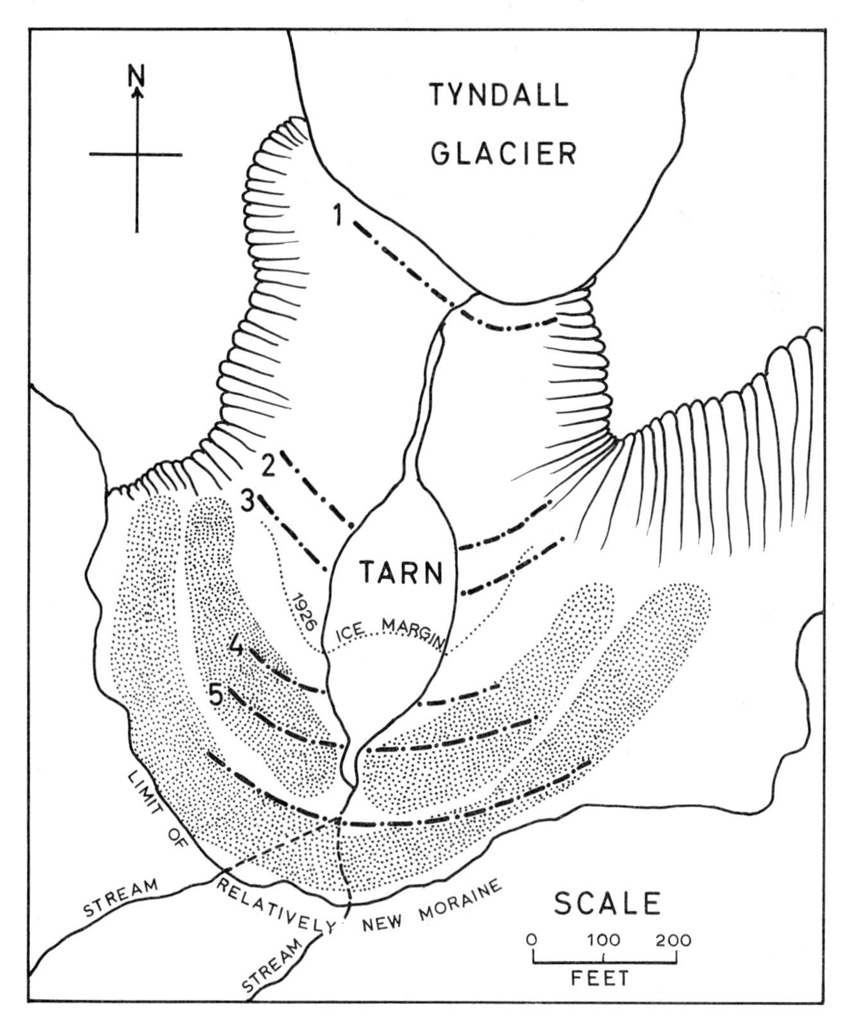

Fig. 15. Tyndall Glacier and Recent Moraines
Order of First Occurrence of Flowering plants 1. Senecio keniophytum 2. Mosses and Lichens and Arabis alpina 3. Agrostis trachyphylla 4. Carex monostachya 5. Lobelia telekii 6. Carduus platyphyllus and Nannoseris schimperi.

retreat is known with some certainty, the moraines of these glaciers are most suitable for such studies.

In the case of the terminal moraines of the Tyndall and Lewis Glaciers, measurements were taken from the point of first occurrence to the ice margin for all species observed. Transects were also drawn through the snout of each glacier recording the frequency of plants across successive terminal moraines. Since these plants occur in small numbers, these latter records proved of little value.

(*a*) *Colonisation of the Tyndall Glacier Moraine.*

The sketch map (Fig. 15) shows the arrangement of moraines at the foot of the glacier. On this diagram the approximate points of first occurrence have been plotted. The first plant to colonise newly exposed moraine is *Senecio keniophytum* (Plate 17), which occurs at a distance of 19 feet from the line of retreating ice. Observations made by glaciologists (CAMERON — personal communication; CHARNLEY 1960) have shown that the Tyndall Glacier is retreating at an average rate of 15 feet per annum. From this figure we can see that this Composite colonises newly exposed moraine within eighteen months of its exposure by the retreating ice. (This fact was re-affirmed when the author returned to this area in June 1960). Although at first sight it seems odd that a flowering plant should be the first to colonise newly exposed moraine, a ready explanation can be found by considering the nature of new moraine. When the ice retreats it leaves behind glacial debris, which consists of smooth, unweathered rocks whose surfaces are quite unsuitable for colonisation by lower plants. At the same time the intense erosion of rocks by the glaciers produces "rock flour" and fine gravel in considerable quantities. Thus at the time of exposure, melt water from above and below the glacier washes both "rock flour" and other small particles into the moraine. It is these patches of fairly fine material that are most easily colonised by flowering plants. Needle ice and other frost agencies cause such patches of fine material to be regularly disturbed, but in the shade of large rocks or around the edges of soil polygons the small *Senecio keniophytum* is able to become established with a great degree of success.

Up to 297 feet from the ice margin (i.e. approximately 20 years' exposure) *Senecio keniophytum* is the sole visible coloniser. The great variability of this plant's growth form (HEDBERG 1957, p. 362) is particularly noticeable in this situation. All the plants in this region are low, decumbent — ascending herbs, densely clothed with long silky hairs which protect both the stem, leaves and the flower buds. The reddish pigmentation is also more noticeable than on plants from lower altitudes of Mount Kenya.

The position of these plants is also very characteristic. In all cases they occur on the protective lee-sided boulders. This is no doubt due chiefly to glacier winds which make the positions facing the ice impossibly cold for plants attempting to colonise. A similar feature is seen on the saddle of Kilimanjaro, where low bushes of *Helichrysum newii* only flower on the lee side, as the cold winds blowing across the area from the peak inhibit flowering on the windward side of the plant.

The first mosses and lichens appear at approximately 297 feet from the ice margin, and apparently occur in no specific order. These plants seem to be able to colonise the moraines when sufficient weathering and frosting has taken place to create a suitable surface for their attachment. Lichens are able to colonise rocks after they have been weathered to a rough surface, or when ridges and cracks have been produced by frost flaking.

Mosses require slightly different conditions. In all sites examined on the moraines they occur almost exclusively on the lee side of rocks, particularly around small patches of frost heaved soil, where they must play an important part in stabilising these materials for subsequent colonisation by flowering plants. Other sites commonly occupied by mosses are found where flakes broken by frost action have not separated from the parent surface and more recently eroded particles have collected in the fissure.

The margin of the Tyndall Tarn provides a suitable, but hardly typical morainic habitat for colonisation by mosses (Plate 18). The milky colour of the water, due to suspended "glacial flour" is particularly clear in this photograph.

Arabis alpina, a short plant with pigmented and hairy leaves, also appears at 297 feet from the ice margin in protected positions. *Agrostis trachyphylla* is found at 330 feet, on the edge of the Tyndall Tarn and *Carex monostachya* at 462 feet, at the lower end of the Tarn. Apparently sterile rosettes of *Lobelia telekii* first occur at 495 feet, and *Carduus platyphyllus* and *Nannoseris schimperi* at 610 feet. In all cases both the frequency of occurrence and also the stature of these plants increases away from the ice boundary. Fig. 16 shows in graphic form the order in which plants colonise the Tyndal glacier moraines.

At 709 feet from the ice margin there is a striking change within a distance of 20 feet from clean, recently exposed rocks to others obviously much older and heavily encrusted with lichens (Plate 19). The former moraine boulders are pale in colour, and less than 10% of their surface is covered by lichens, while that of the latter is blackened by a very dark lichen that covers 90% of the surface area. *Usnea* becomes attached in its turn to these flat spreading lichens, in the form of short tufts up to six inches long.

At the point of sudden change, *Anthoxanthum nivale, Helichrysum citrispinum* var. *armatum, Pentaschistis minor* and *Senecio kenioden-*

dron (small and seldom exceeding 4′) enter the association within a few feet and soon become quite common.

Assuming an approximately even retreat of 15 feet per annum in recent years, it seems that the edge of the newer moraine must be under 100 years old. If the rocks constituting these moraines have taken this length of time to become colonised by lichens to approximately 10% cover, the blackened well covered area must have been exposed from the ice for several hundred years. When the

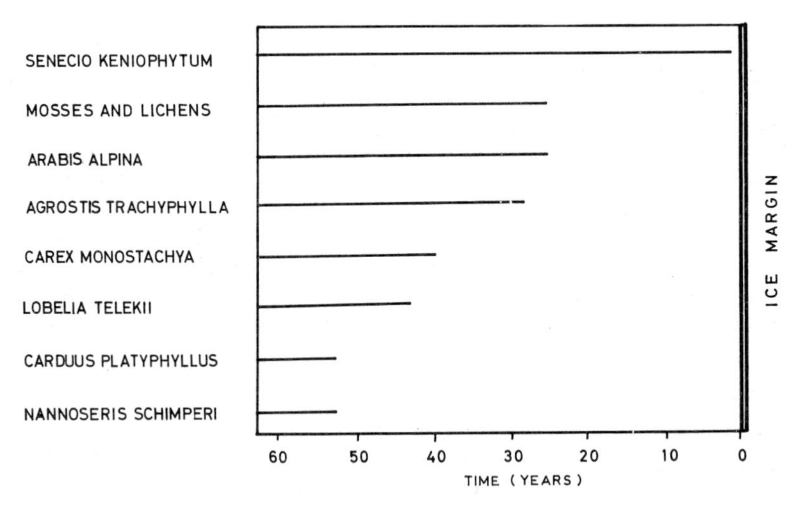

Fig. 16. Rate of Colonisation of Tyndall Glacier Moraine.

phenomenon was first observed in January 1958, the most likely explanation seemed to be that the glacier remained stable for many years but began a new line of retreat in about 1860. Although this explanation is feasible, it cannot in any circumstances be said to explain the sharp line of demarcation and close approximation of the new and the ancient moraines. In 1960 the author returned to the glacial moraines and, after re-examining the area came to the following conclusions:

1. If the glacier had remained in one position for some considerable time (in fact, for several hundred years) there should be visible an area of graduation of colonised rocks. This is not so, for the line of change makes a sudden, accurate crescent around the obviously recent moraine.

2. If we discard this idea, then the alternative would be to postulate that the glaciers almost, or completely disappeared in the comparatively recent past, and then entered a short period of ad-

vance. In these circumstances, with the peaks completely deglaciated, rocks would have become weathered and colonised at a much greater speed than is the case when ice is present. As soon as the ice began to advance again, it would have ground those rocks in the glacier's path and, at its point of lowest descent, have left a distinct crescent shaped moraine overlaying well weathered and colonised rocks of obviously much greater age. This hypothesis fits the observable facts well on Mount Kenya where this boundary is so distinct, and in addition, when the author examined the terminal moraine of the Tyndal Glacier, the recent moraine was seen apparently to overlay older material.

HEINZELIN (1953) (pp. 12—14) supports this theory with regard to Ruwenzori, and he even correlates this short period of advance on the Equatorial mountains with the "Wurm" or "Little glaciation" of the Alps. If, as he suggests, these small glaciers reached their maxima between 100 and 300 years ago, our calculations of glacier retreat and rate of colonisation would appear to be correct. GREGORY (1894) (pp. 521—522) talks of the Lewis Glacier breaking through one of its terminal moraines, when he visited it in 1894, and also states that he found the remains of dead *Lobelia* stems on the moraine. This is a valuable observation, for since no *Lobelia* plants are found in flower at this altitude, one can only assume that in the recent past conditions at this level must have been more favourable for their growth, and that a new period of glacial advance once more depressed the upper limit of their range. If this observation is accepted, it provides a useful piece of evidence to correlate a re-advance of the glaciers on Mount Kenya with the Little Ice Age of Europe.

The marked change from barely colonised recent moraine to well colonised old moraine can be seen below all the main glaciers, but since most of them are retreating over very steep ground, or are retreating along an uneven line, time succession studies are more difficult.

(*b*) *Colonisation of the Lewis Glacier Moraine.*

Although the Lewis Glacier provides some excellent terminal moraines for colonisation studies, the region is not quite so perfect from a time-succession point of view as is the area below the Tyndall Glacier. Over the last forty years the Lewis Glacier has retreated behind a large cliff which now holds a proglacial tarn; for this reason one must suppose that plants growing on the moraines are more protected from the cold glacier winds than their counterparts on the Tyndall moraine (Fig. 17). It will be noted that in fact the rate of colonisation has a slightly different time scale, which is in part due to the fact that the Lewis Glacier is retreating at a faster rate than the Tyndall Glacier. Calculations made by the I.G.Y.

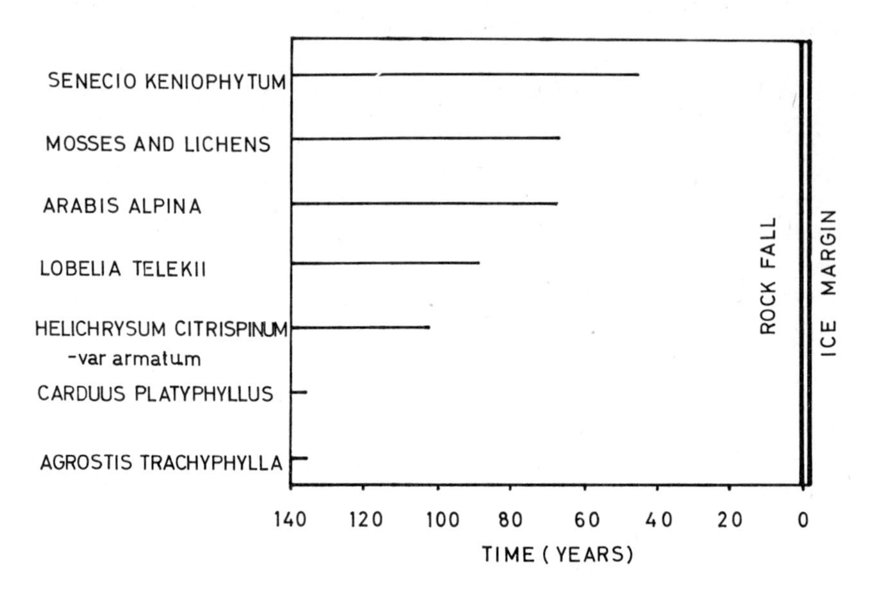

Fig. 17. Rate of colonisation of Lewis Glacier Moraine.

Mount Kenya Expedition agree that this glacier is retreating at a rate of between 25 feet and 30 feet per annum (Fig. 18).

The order of colonisation on the Lewis moraine bears many similarities to that of the Tyndall. *Senecio keniophytum* is in the forefront of succession, followed after an interval of 231 feet by mosses, lichens and *Arabis alpina*. Other plants follow in the order: *Lobelia telekii, Helichrysum citrispinum* var. *armatum, Carduus platyphyllus* and *Agrostis trachyphylla*.

It is difficult to account for the presence of *Helichrysum citrispinum* below the Lewis Glacier while it is absent from the Tyndall. It appears, however, to be a hardy plant, well adapted to cold nights, hot days and a very dry atmosphere. The explanation may be found in the extra protection from cold winds afforded by the rock fall below the Lewis Glacier.

In the foregoing discussion, the marked change from the smooth, little colonised rocks of the new moraine to the well weathered material of the older moraine was noted with regard to the Tyndall glacier. The same change is found on the Lewis moraines, the sudden change also taking place within a distance of about 20 feet. The total rock surface covered by mosses and lichens in this region is again 90%. This region is, in fact, continuous with the same region below the Lewis Glacier.

(c) Proglacial Tarns.

Both the Lewis and the Tyndall Glaciers hold, or have held at their snouts in the last 40 years, small tarns. The Tarn at the foot of the Lewis Glacier is still held by the snout, and is enclosed at its lower edge by the rock fall. The Tyndall glacier Tarn is no longer

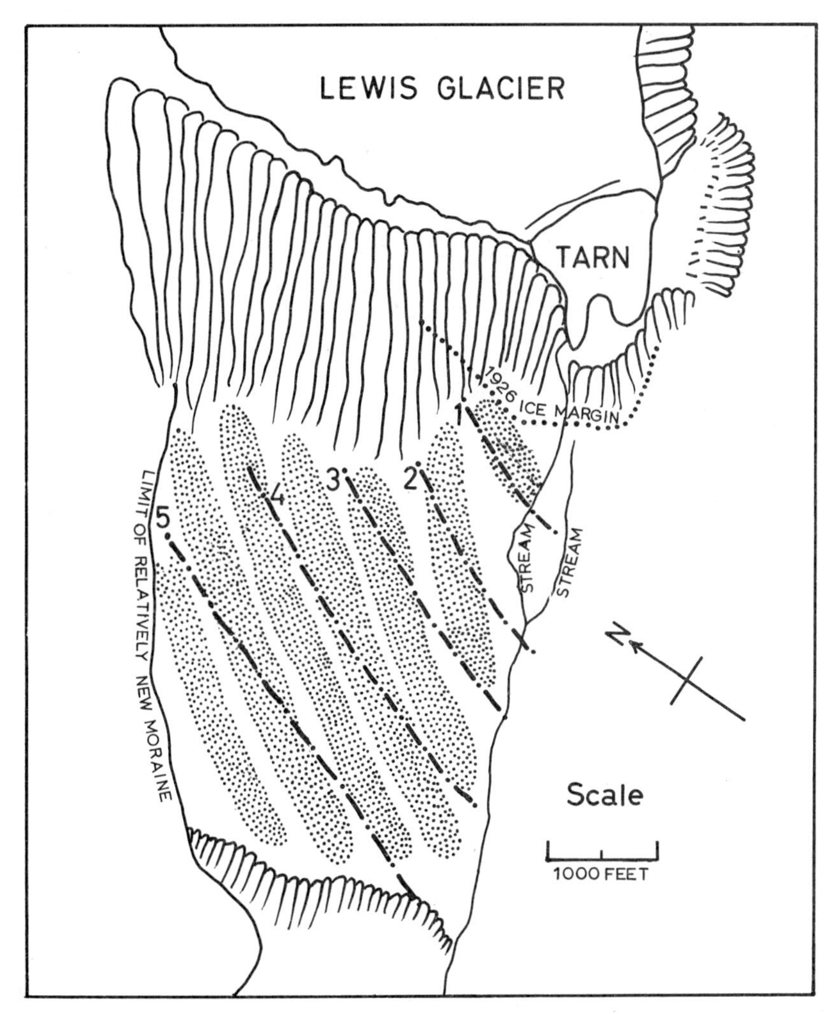

Fig. 18. Lewis Glacier and Recent Moraines
Order of First Occurrence of Flowering Plants 1. Senecio keniophytum 2. Arabis alpina 3. Lobelia telekii 4. Helichrysum citrispinum var. armatum 5. Carduus platyphyllus and Agrostis trachyphylla.

in contact with the retreating ice, although in 1926 (DUTTON 1929, facing p. 94) it was held by the snout and limited at its lower edge by the terminal moraine.

The Tyndall Tarn has an opaque appearance due to large quantities of glacial flour suspended in it (Plate 18). Although the temperature of this water seldom exceeds 3 °C, it supports a rich growth of algae and an active community of Turbellerians, Copepods, Ostracods and freshwater mites. The Lewis Glacier Tarn is usually covered with ice and the only living organisms found at the foot of this glacier are algae, which occur in small pools on the top of the rock fall.

(d) Dating Moraine Deposits.

By comparing photographs taken at the beginning of this century with surveyed positions of the present snout, it is possible to determine the approximate rate of retreat. The balance between the amount of new ice being formed each year, and the old ice melting away, is one which is largely dependent on precipitation in any one season. The unusually heavy rains of 1961 and devastating flooding may well have produced a heavy ablution season that may later be reflected in a slight advance of the glaciers.

The following are the figures computed by BAKER (1959) giving the average retreat rates for the Mount Kenya glaciers:

Glacier	Distance below terminus at January 1958	Year for which comparison computed	Average rate of retreat per annum
	Feet		Feet
Lewis:	C. 1500 *	1893	23
	C. 1100 *	1926	34.4
	700 †	1934	29.7
	500 *	1938	25
	400 *	1947	36.4
	220 †	1950	27.8
Tyndall:	107 †	1950	13.5

* Figure estimated by comparison of photographs and sketch maps.
† Figure obtained by survey measurement (F.E. CHARNLEY, personal communication).

It is interesting to note that the glaciers for which vegetation colonisation studies were made show the maximum rate of retreat. Other glaciers around the peaks show little or no recession. (Darwin glacier 7.07 feet per annum, Krapf glacier 3.8 feet per annum. The retreat of the Cesar, Joseph and Northey glaciers was too small and steep to be measured). This may be attributed to their aspect and the actual cliff faces on which they occur.

It has already been shown that with a known approximate rate of retreat, it is possible to determine the rate at which plants colonise newly exposed moraines. These figures are significant for deposits of similar aspect and altitude, but it is the opinion of the author that in marginal environments such as the Nival Zone, it is impossible to compute rates of colonisation that can be applied to all regions, and still less to other mountains.

Numerous attempts have been made in the past to produce a yardstick by which rates of colonisation may he measured. HAUMAN (1935) calculated the rate of growth of a megaphytic *Senecio* sp. on the western side of the Ruwenzori range. These figures have been quoted by other authors as indicative of *Senecio* growth in general (WITTOW 1959). Such measurements of growth rate are only applicable to, and can only be used for plants in a particular area, or plants submitted to similar microclimatic conditions. On Mount Kenya *Senecio keniodendron* shows a considerable variation in rate of growth between the top and the bottom of steep valley slopes; such calculations are of no use on Mount Kenya and indeed, it is suspected that they are of little application elsewhere, except possibly in the region where the original figures were calculated.

Lichens have also been used in determining the rate of colonisation of newly exposed rocks. Here again great caution must be used. On the newly exposed moraine of the Tyndal and Lewis glaciers, the diameter of lichen colonies shows considerable variation, depending apparently upon the degree of shade, the height from ground level and the aspect of the colony. No doubt with observation carried out over a long period, it would be possible to calculate a time/growth factor, but with the present state of our knowledge of lichen growth, it is not an accurate means of determination. A rough, but useful figure can be obtained by calculating the percentage of rock surface covered by lichens and mosses and relating this to known retreat rates.

The Nival Zone of the Equatorial mountains provides an ideal, if uncomfortable, habitat for studying colonisation, and it is hoped that in the near furute it will be possible to carry out similar studies on other mountains in East Africa.

2. Other Phases of Colonisation.

The factors contributing to the distribution of newly formed soils have already been mentioned, and it is these surfaces that play a vital part in determining the types of vegetation that may become established upon them. Throughout the Alpine Zone, frost is an important soil transport agency, and in many exposed situations it has been shown that colonisation by plants may be completely repressed.

In the Upper Alpine Zone the steep slopes of valley walls and valley heads lead to large areas of unstable ground. These are usually scree slopes, on which the action of gravity itself is an important sorting agency, and pebble glaciers which contain no fine fractions are hence devoid of vegetation. In fact, these mobile surfaces frequently cover and destroy areas that are already well colonised; (e.g. the head of the Nairobi Valley: 14,000 feet). Mobile scree surfaces do not provide a material that is suitable for colonisation, but they may often be stabilised from their more slowly moving edges. This phenomenon can frequently be observed in the region of the peaks, where narrow strips of scree form below narrow cliff cracks. In these situations *Veronica gunae* has been found creeping over the edges of loose mobile scree, rooting at the nodes and very effectively binding the particles together. At the outer edge of this binding agent, small tussocks and mosses had become secondarily established (e.g. scree at the head of the Nairobi Valley). *Galium* sp., although not in flower, formed extensive patches of vegetation, binding the edges of narrow bands of scree which descended from water worn cliff cracks. In this situation little other than mosses had become established. Where ledges collect a little soil, the most common plants to be found binding them in place are mosses, *Galium glaciale, Myosotis keniensis, Arabis alpina, Senecio keniophytum* and *Senecio purtschelleri*.

Where soil has been carried to flat ground at over 14,000 feet, the most important and probably the first coloniser seems to be tussock grass. When these plants have become established, the typical damp ground associates, such as *Haplocarpha rueppellii* and *Ranunculus oreophytus* first become established around the tussock bases. Since the ground between the tussocks is frequently disturbed by needle ice, any further colonisation of this ground seems to be prevented until mosses have stabilised the soil.

It seems probable that vegetation has never become established on ridge tops in the Upper Alpine Zone, due to extensive frost heaving. On valley walls leading from these ridges, *Senecio keniophytum* and *Alchemilla argyrophylla* ssp. *argyrophylla* become established in a very dense stand, due to the abundant availability of surface water. These stands, however, appear to be slowly depressed down the wall owing to the rapid rate of removal of the smaller soil particles. The phenomenon of alternative bands of *Senecio keniodendron* and *S. brassica* has already been attributed to the transport of fans of fine material to valley bottoms by meandering streams. On steep cliff faces, where fine material becomes evenly spread along the base (e.g. Gorges Valley) the primary cover seems to be dense tussock grass, which later becomes eroded in places by stream courses, and these areas become secondarily colonised by bands of *Senecio keniodendron.*

Since the main parent material of Alpine soils on Mount Kenya is morainic, these deposits show perhaps better than any other a graduation of plant community development. This gradual development is directly related to the degree of erosion of the moraine deposit until finally it almost entirely loses its identity.

(*a*) *New Moraines:* (Tyndall and Lewis Glaciers).

These have already been described and they are characterised by being very little colonised and by containing a great deal of fine material in their soils. They are colonised first by *Senecio keniophytum, Arabis alpina* and *Agrostis trachyphylla.* Later scattered plants of *Nannoseris schimperi, Carduus platyphyllus* and *Helichrysum citrispinum* var. *armatum* enter the association, together with apparently sterile specimens of *Lobelia telekii* and *Senecio keniodendron.*

(*b*) *Moraines of Intermediate Age:* ($\pm$ 1,000 years).

These deposits are found in the heads of most valleys (13,000—14,000 feet) or below the higher Alpine lakes. Their surfaces are well weathered and up to 90% of the surfaces area of the boulders comprising them are covered with lichens. The material between them is coarse, except in pockets where humus collects. They all carry well developed stands of *S. keniodendron* and on flat ground, *Senecio keniophytum, S. purtschelleri* and *Arabis alpina* are well established, particularly in the moraines that contain Hyrax colonies. *Alchemilla argyrophylla, Helichrysum cymosum* and *H. citrispinum* are to be found on the sides, together with small clumps of *Blaeria filago.* Small patches of fine soil between boulders carry *Carduus platyphyllus, Myosotis keniensis, Nannoseris schimperi* and *Valeriana kilimandscharica* ssp. *kilimandscharica.*

In almost all situations other than rock strewn valley walls it seems likely that the primary cover is tussock grass, which forms dense mats in places. Grasses are usually not well developed, forming small tussocks no more than one foot in height.

(*c*) *Well Weathered Historic Moraines at Lower Altitudes:* (Gorges Valley, 11,750 feet)

The vegetation of these moraines has already been described and it has been said that they represent one of the later stages of weathering of an Alpine moraine. They have lost much of their original identity, and the raised moraine deposits now consist of little more than crescentic heaps of boulders, with little soil

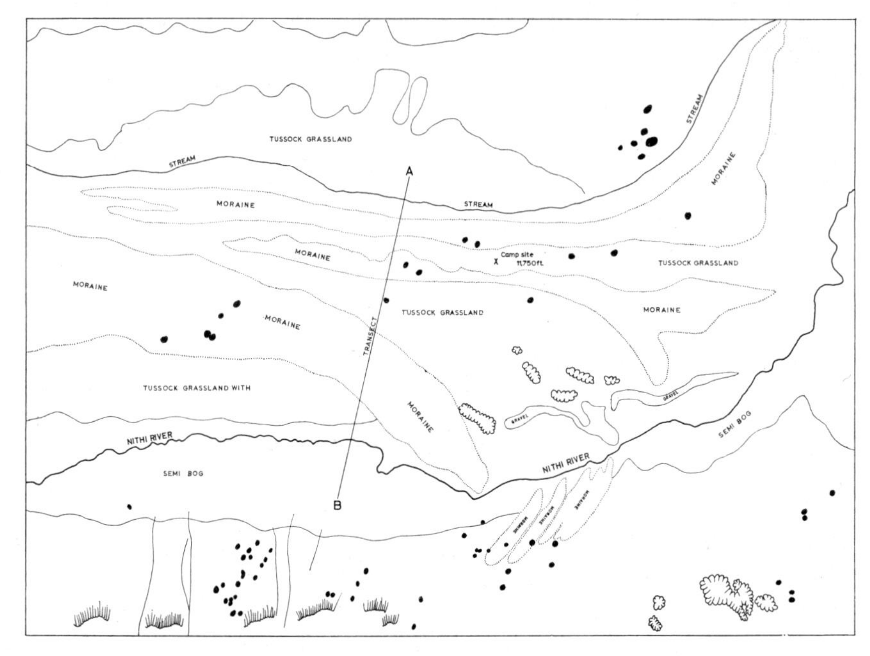

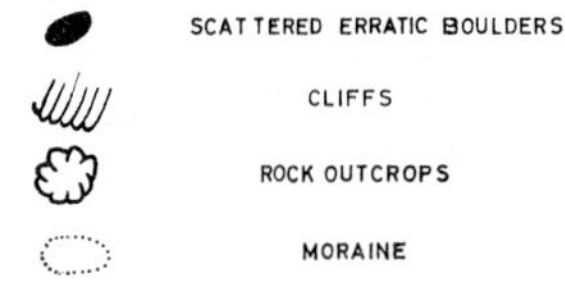

Fig. 19. Historical Moraines at Foot of Gorges Valley.

between them (Fig. 19). Much of the smaller plant life has been replaced by woody shrubs such as *Protea kilimandscharica* and the Heaths, although small aggregations of humus below these shrubs support a certain number of small herbs. The inter-moraine areas are composed of fine deep soil, derived for the most part from fine fractions of the adjacent moraines and a certain degree of material accumulated from ridges higher on the mountain. These grassland tussock areas contain some scattered herbs and have been secondarily eroded by small streams. While these three series of moraines show a gradual change of vegetation which may be related to their progressive weathering, it must be remembered that as the glaciers have rededed, so the climate in the lower areas has become less extreme, and this must also play an important part in their accelerated colonisation. Fig. 20 shown a typical transect drawn across these moraines.

At the present state of our knowledge of the climate of Mount Kenya, this is as far as we can go with regard to tracing the history of colonisation of the Alpine Zone.

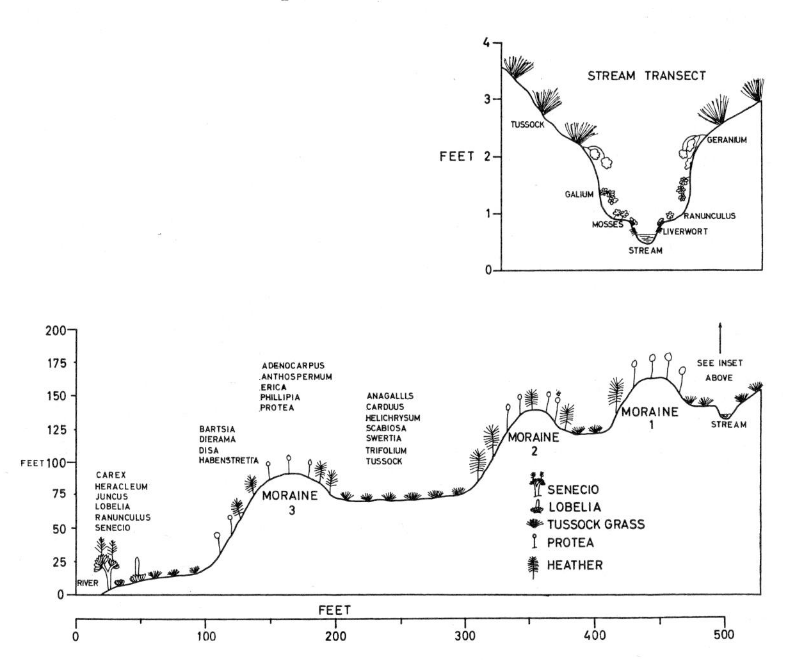

Fig. 20. Transect across Historical Moraines at the Foot of the Gorges Valley.

BIOTIC FACTORS IN THE ALPINE ZONE

Animals as well as plants are subjected to the peculiarities of a high altitude climate on Mount Kenya, just as they are on all the other tropical mountains of the world. However, unlike the plants, they are able to remain in a niche wherein they are protected from the environmental extremes to which the vegetation is exposed. Our knowledge of the behaviour of animals in these niches is very poor with regard to vertebrates, but more complete when we consider the invertebrate fauna. MANI (1962) has considered this question in some detail in his "Introduction to High Altitude Entomology". He found, working on Himalaya, that around a small boulder niche there was a great variation in microclimate. The data recorded at this site is given below:

Situation	Temperature	Relative Humidity
Atmosphere	-1.3 °C	40%
5 cm above rock surface	26 °C	12%
Rock surface	30 °C	
Under rocks in insect niche	10 °C	98%
Fissure in rock	18 °C	70%

These figures clearly show the importance of both temperature and humidity to insect survival. In bright sunshine (as above) the relative humidity falls rapidly, and an Arthropod that remains exposed becomes rapidly dessicated. The author has observed the same phenomenon on Mount Kenya (see page 64), where he found that at ground level the relative humidity just before sunrise was 90% and that when the sun rose at 7,50 a.m. this figure fell in about 90 mins. to below 20%. When during the course of the day the sun became obscured, the figure rose rapidly to 80%. Such sudden changes in relative humidity are obviously of great importance to an insect, greater even than rapid changes in temperature. The great importance of such physical factors is expressed by the large number of apterous insects found in the Alpine zones of these mountains. SALT (1954) has described and figured them in his description of the invertebrate fauna of upper Kilimanjaro, and MANI (1962) has commented on the same feature on Himalaya.

The insect fauna of Mount Kenya is very little known, but the author has observed flightless Acridids, similar to the *Parasphaena pulchripes* recorded on Kilimanjaro by SALT, and no doubt many more exist. Even insects that have not lost the power of flight live a very sedentary life and seldom emerge from the protection of their individual niches.

The Bibionid (Diptera) flies that live in the giant *Lobelia keniensis* flowers enact their whole life cycle in this position, and adults, eggs, larvae and pupae can be found in the same flower. The wings, however, of this fly are not appreciably reduced, and they form the most important item of food of the Scarlet Tufted Malachite Sunbird, since they make little or no attempt to escape the predations of this bird. *Simulium dentulosum* form *macabae*, which breeds in very large numbers in streams almost up to the foot of the glaciers, provides another interesting example of adaptation to a high altitude existence. In spite of the large numbers of larvae and pupae to be found in streams, an intense search for the adults on vegetation, under stream banks, and in Hyrax and Otomys burrows, only produced three adults in several hours, although adults were seen emerging from fast flowing water. It can only be presumed that these insects migrate to lower levels after emergence, and later return to breed.

The swarms of Chironomids that are seen rising and falling over *Lobelia keniensis* rosettes, where they breed, disappear at night or when the sun is covered by cloud. These insects do not seem to migrate, but shelter between the persistent leafy frills of the Giant *Senecio keniodendron*.

Species of Beetles and Spiders that inhabit the Alpine Zone shelter under stones or vegetation and are only seen when the sun is warm (not hot) and the humidity is about 60%. The same is true of molluscs that are largely found under stones and in damp vegetation, such as the spaces provided between the leaves of the Giant *Senecio* and *Lobelia*.

The difficulties confronting the other insects in surviving in the Alpine Zone are well illustrated by those found in the region of the glaciers. Up-currents of air are continually carrying debris from low levels, to be deposited around the peaks. ZEUNER made a study of these wind-blown Aeolian deposits in 1945. These materials include spores, pollen, seeds, grasses, inorganic and organic dust. Included in this debris are quite a large number of insects. Although the author has never observed the large aggregations of insects carried up and deposited on the ice, as has been observed in the Alps, in the Himalayas and other high mountains, lowland insects are, however, quite common, Aphids and Diptera being frequently found on the surface of the ice. Butterflies, being perhaps the most conspicuous of the insects so transported, are commonly seen on the Lewis and Tyndall glaciers. During 1958, while working on the Lewis glacier in bright sunlight, the author observed small holes in the ice, up to 6″ deep, and at the bottom of each hole a Lepidopteran was recovered. At the same time butterflies were seen flying above the glacier. Since these are never observed in the early morning, it seems likely that these insects are carried up in currents of air during the day and at night they die from the intense cold and fall into the ice. When the sun rises next morning, the body and wings absorb a certain amount of heat which melts the ice below the insect's body so that it slowly sinks into a small pit. The following species were recovered, either from the surface of the ice or from small pits:

Colias alecto LIN. (3 male, 2 female)
Harpendireus aequatorialis SHARPE (1 female)
Vanessa cardui LIN. (1 male)
Zizula hylax F. (1 male)

Although such phenomena can be paralleled in other parts of the world, the peculiarities of diurnal climate which obtain on Equatorial mountains create conditions which would make a protracted entomological study in these regions very rewarding (Cf. MANI, 1967).

1. Relations Between Animals, Vegetation and Habitats.

Faunal observations were confined largely to the Vertebrates on the mountain, and although in many ways they are better protected against extremes of climate than the invertebrate fauna, they are undoubtedly subjected to a marginal existence. The most noticeable features of the vertebrate fauna on Mount Kenya is the small number of species that are resident in the Alpine Zone, together with the locally large number of individuals of these few species. Although it is a well-known feature on other high mountains, it is of interest to confirm that all these mammals have unusually long fur, that of the Hyrax being over two inches long on Mount Kenya, while species of the same genus (*Procavia*) living in the lowland savannah have fur no longer than half an inch. The Hyrax also follows BERGMANN'S rule in that they are probably the largest Hyrax to be found in Africa.

Members of the Queen Elizabeth College, London expedition to Mount Kenya held between January and February 1965 measured temperatures inside Hyrax burrows and compared them with recordings made in the surrounding air. That of the burrow showed a mean maximum over four days of 9 °C and a mean minimum of 0.6 °C, while that of the air outside over the same period displayed a mean maximum of 9 °C and a mean minimum of —4 °C (SALE 1965). Thus the temperature in the burrow remained above freezing albeit fractionally. In this connection it should be noted that the Hyrax lives in holes between rocks, which are seldom deep, and except for perhaps scraping out litter or other debris, is in no sense a burrowing animal.

A vegetation which is submitted to great climatic extremes

Plate 19. Margin of old and recent moraine below the Tyndall Glacier: 14,500. feet

Plate 20A. *Procavia johnstoni mackinderi* in the Mackinder Valley: 14,000 feet.

Plate 20B. Rock Fall at the entrance to the Gorges Valley. One of the largest Hyrax colonies on Mount Kenya.

shows a marked morphological adaptation to its environment, and except where fire intervenes in the Northeast and North-west, the plant associations are relatively slow growing. This means that a large number of herbivorous mammals would exert a profound feeding pressure on the vegetation, even presuming that a large percentage of the plants available are utilised as food. Besides the obviously slow replacement of vegetation, many of the plants have developed protective devices against the large diurnal range of temperature and intense radiation. These modifications mean that, except in their very young stages, they are not usually utilised as food by herbivores. In addition to variations in growth form, morphological modifications are extremely common; hairy or tomentose leaves, reduced leaves, spines or coriaceous surfaces all render plants inedible except in severe seasons, when almost anything is consumed by herbivores. (SALE, 1961 in a personal communication on his observations on Hyrax in the Rift Valley, states that in severe drought the animals began to eat bark and even wood when all available green material has died down).

The most direct pressure exerted on the vegetation in the Alpine Zone is due mainly to herbivorous vertebrates, while at the same time, extreme physical factors create a pressure on both vegetation and herbivores by virtue of the modifications they produce in the plants that render them unavailable as normal items of a vegetable diet.

The most important herbivores in the Alpine Zone cycle of animal-vegetation balance are the Hyrax *(Procavia johnstoni mackinderi)*, the Groove-toothed Rat *(Otomys orestes orestes)* and the Common Duiker *(Sylvicapra grimmia altivallis)*. Before considering their effect on the vegetation, it is of interest to look at the habitats occupied by these creatures and to formulate as far as is possible the inter-relations of their respective communities. Four situations where all three animals live have been studied, these are: (a) Mackinder's Camp at the head of the Teleki Valley (13,750 feet), (b) Lake Höhnel (13,700 feet), (c) Hall Tarns (14,000 feet), and (d) Mackinder Valley (13,700 feet). All these stations comprise old moraine deposits or weathered phonolitic crags bordering open tussock grassland, and are close to water. The Hyrax occupy the rocky habitat which has been sufficiently weathered and eroded for them to become established. Below in the tussock grassland Otomys Rats live in burrows they excavate at the bases of Giant Senecios or tussocks. The small numbers of Duiker that occupy most valleys are found in *Senecio keniodendron* forest with a more or less dense ground cover of *Alchemilla argyrophylla*. Mingling of the three communities takes place only when the Hyrax and Duiker leave their own habitat to go to water or when foraging in the Otomys habitat of the valley floor, otherwise the communities of these three dominant herbivores are conspicuously isolated from one another by the specific nature of the niche they occupy.

We now have records of the Harsh Furred Mouse (*Lophuromys a. aquilus* TRUE), the Giant Mole Rat (*Tachyoryctes rex* HELLER) occurring up to 13,000 feet. The former is a somewhat omnivorous Rodent which takes a variety of vegetable matter and insects in its diet, while the latter is a burrowing Rodent that feeds largely on roots and tubers, but will often feed on the aerial parts of the plant by pulling them below ground before eating them. There also seems little doubt that at least one Shrew (probably *Crocidura alex alpina* HELLER) also enters the Alpine habitat. The ecological separation of these species is not yet completely apparent, but work is currently in progress in this field.

At the time of writing (February 1966) the author is investigating the dry Northern slopes of the mountain, where a large number of plants and vertebrates appear to extend their range higher than on any other segment. Perhaps the most striking example being an apparently resident herd of Zebra at 13,500 feet (apparently *Equus burchelli* GRAY).

In addition to this apparent habitat isolation, there appears to exist amongst the Alpine vertebrates what might be called a "food spectrum". HUXLEY (1961) speaks of a similar situation in savannah herbivores where he writes "In the African bush or savannah the herbivores show an astonishing spectrum of structure and habitat, an ecological division of labour which leads to optimum utilisation of the vegetable resources". Such a condition will be produced in an environment where, due to scarcity of food, the animals evolve dietary patterns which only overlap each other slightly and hence eliminate to a large degree inter-specific competition for nutrients. By examination of stomach contents, droppings and cropped plants in the region of burrows, the following information has been obtained:

Procavia johnstoni mackinderi THOMAS. (Plate 20 and 21) (Mount Kenya Hyrax) 12,000 feet to 15,500 feet.

Analyses of stomach contents of animals from the Teleki Valley and Gorges Valley show that the animals are selective in their feeding habitats, with a preference for grasses and mosses. These, however, varied with the position of the colony; five specimens living in ridge top colonies of the Teleki Valley contained a greater percentage of grass than those in valley bottoms (e.g. Mackinder and Hall Tarns sites) where, due to close proximity to permanent water, moss is a more dominant constituent of the vegetation.

Reliable information was obtained by examining plants in the vicinity of the burrows. This showed that on raised solifluction terraces, young plants of *Festuca, Agrostis* and *Pentaschistis* were taken; while along streams and lake sides *Subularia monticola, Crassula granvikii, Ranunculus oreophytus* and *Oreophyton falcatum* were common items of diet. Young *Lobelia keniensis* rosettes were sometimes nearly cropped of their leaves although older plants were seldom touched.

Colonies at the head of the Mackinder Valley surrounded by *Senecio keniodendron — Alchemilla argyrophylla* forest eat a great deal of the leaves of the latter as an apparently normal item of their diet.

Although lowland Hyrax do not seem to need to drink, the Mount Kenya animals are often seen drinking at streams, pools or lakes, their feet often forming deeply worn tracks from their burrows to the water's edge. These paths are often over 100 yards long, quite straight and noticeably without side tracks, which suggests that they do not appear to feed in the coarse tussock grassland between their burrows and the water. Undoubtedly a well worn track without side runs has a decided element of safety for the animals when they are caught in the open. The author has seen a Hyrax taken from one of these tracks by an Augur Buzzard in full flight towards its colony.

Many plants which grow in the immediate vicinity of the burrow entrances are never eaten. These include *Heracleum elgonense, Valeriana kilimandscharica* ssp. *kilimandscharica, Senecio brassica, S. keniodendron* and *S. purtschelleri*. In addition, the succulent leaved *Sedum ruwenzoriense,* which grows profusely on the deposits of Hyrax dung that collect on rocky ledges, also is seldom eaten. All these plants, except the last named, are either coarse or aromatic in foliage. It would seem that, since these plants very effectively camouflage the entrances to burrows, the fact that they are not eaten is of mutual advantage to both mammal and the survival of the plant (Coe 1962).

Otomys orestes orestes Thomas. (Groove-toothed Rat). 11,000 feet to 14,000 feet.

These long furred rodents are found in large numbers in the Alpine Zone, particularly in tussock grassland, on valley floors or at lake-sides. Almost every *Senecio keniodendron* in the area has a small hole at the base of its trunk made by these rodents. The burrow leading from this entrance does not go down into the soil, but leads up into a small cavity excavated by the creatures at the bases of the persistent leaves, where the temperature is reasonably constant. In these grasslands the runs form maze-like patterns between the tussock bases. Stomach contents of these small mammals were not very helpful as the vegetable material consisted almost exclusively of the fibrous remains of plant roots, mixed with a smaller percentage of seeds and aerial shoots that were chewed to tiny fragments. From watching the animals on the mountain one learned that they emerged from their burrows and foraged along the sides of their runs, taking seeds and a little grass. The runs invariably led to water or swampy ground, and many burrows were actually in the beds of streams, animals on several occasions being seen to swim back to their holes. Around the edge of the water were specimens of *Subularia monticola* and *Crassula granvikii*, chewed but not eaten to any degree. The commonest item of food on the waterside appeared to be detritus that had been washed ashore.

Sylvicapra grimmia altivallis HELLER. (Common Duiker). Probably from the forest edge to 14,000 feet.

Although this animal is another important herbivore, it cannot compare in numbers with the population of the two preceding species of mammals. After extensive travelling in and around the Teleki Valley, an optimistic estimate of their total population in this area would be six pairs. They are extremely shy animals and remain for the most part concealed within the *Senecio* — *Alchemilla* forest, or at lower altitudes in patches of *Erica* bush. This Duiker is chiefly a browser, and has been seen eating *Alchemilla* and a selection of other small woody plants that grow in the region. In contrast to other members of the *grimmia* species, they feed mostly in the daytime, lying up at night; and they seldom visit water, but when they do, they often eat the young rosettes of *Lobelia keniensis*.

This information shows that, not only is there a distinct spectrum of habitat exhibited by these herbivores, but their food preferences also tend to isolate them from inter-specific competition. Such division in an environment where food is nowhere abundant would seem to be the most effective mechanism for controlling and preserving the natural vegetation. The idea of segregated food preference, or a spectrum, may be extended further by considering the other resident vertebrates of the Alpine Zone.

Nectarinia johnstoni johnstoni SHELLEY (Plate 22) (Scarlet Tufted Malachite Sunbird). 10,000 feet to 15,000 feet.

This Sunbird is undoubtedly the most striking bird in the Alpine Zone, and although their numbers cannot be compared with *Procavia* or *Otomys*, they are common from the edge of the forest (10,000 feet) to the foot of the main peak (15,000 feet). Non-breeding parties aggregate on the moorland, showing a preference for regions where well-established stands of *Protea* exist. In the Alpine Zone, where it seems likely that they breed throughout the year, they are territorial in their behaviour, the size of the territory varying with the altitude, and in the consequent availability of food. Hence the population is denser in the lower Upper Alpine Zone than in the peak region (Coe 1960). The adults are insectivorous and for the most part they feed hovering in front of inflorescences of *Lobelia,* which contain large numbers of insects. Stomach contents of these birds show that about 90% of their food is composed of Bibionid Diptera which both feed and breed in the bases of *Lobelia* flowers. Insect collections made from these inflorescences in different valleys has shown that between 90% and 95% of those present are Diptera. Another regular source of food for these birds are the Chironomids that breed in small pools of water held by the rosettes of *Lobelia keniensis.* These flies hover over the rosettes in small clouds and the Sunbirds sweep through the swarm and catch them after the manner of a "fly-catcher". In fine weather when the birds are nesting, they feed their young almost exclusively on the small Lycaenid butterfly *Harpendireus aequatorialis* (Coe 1960).

Pinarochroa sordida ernesti Sharpe. (Plate 23)
(Hill Chat). 10,000 feet to 15,000 feet.

The Hill Chat is a very common little bird throughout the Moorland and Alpine zones. They are very friendly birds and quite fearless, actually hopping inside a tent to collect scraps. Although these birds will take almost anything that is offered, their stomach contents indicate that their main diet is small Carabid beetles, weevils (Curculionidae) and spiders. The food is collected both on the ground and from between the leaves of the Giant *Senecio* and *Lobelia.* Hill Chats, however, never feed on the *Lobelia* inflorescences visited by the Scarlet Tufted Malachite Sunbird.

Serinus striolatus striolatus.
(Streaky Seed Eater). From lowlands up to 14,000 feet.

This bird is locally common in the Alpine Zone. The stomach contents show a high percentage of seeds mixed with a quantity of vegetable detritus. It was not possible to identify seeds from stomach dissections, but it would appear from watching these birds that their diet is almost exclusively seeds in the alpine zone. They were observed frequently eating seeds from the rosettes of *Nannoseris schimperi* and *Carduus platyphyllus.* The erect seed capsules of *Romulea keniensis* are also frequently visited.

These three species of birds are by far the most common avian residents of the region. The following occur in small numbers in each valley, and while they do not exert any profound effect on the habitat, they illustrate again the existence of a segregated food preference.

Apus melba africanus (Temminck)
(Alpine Swift). Widespread, up to 16,000 feet.

These birds are undoubtedly resident in the Alpine Zone and have

been recorded breeding in October. The author found disused nests in the Gorges Valley (13,000 feet), under the Hall Tarns cliffs, in January 1958.

Several groups of about thirty birds seem to live around the mountain, where they move up and down the major valleys ranging between the moorland and the lakes and glaciers of the peak area. Food consists largely of insects which they catch along stream courses and over lakes and tarns. In January 1960 the author observed a group of birds over the Tyndal glacier and Two Tarn, where they were feeding on Diptera and Lepidoptera carried to the peak area with wind blown debris.

Anas sparsa leucostigma (RUPP.)
(Black Duck). Found throughout the forest and Alpine zone.

An essentially highland bird, resident on the streams and Tarns of the Alpine zone. Although they do often fly up and down the valleys, many birds remain resident in the region for days or even weeks. (Lake Höhnel January 1960). It has been recorded breeding on Mount Kenya at Lake Höhnel, the Hall Tarns, Lake Teleki and Enchanted Lakes.

Food in the Alpine zone consists largely of submerged plants and stream or lakeside debris. On Lake Höhnel the apparently sterile *Potamageton (schweinfurthii?)* is eaten in considerable quantities, and after the plants have been pulled and chewed, the birds feed further on the detached leaves when they are washed ashore. On other lakes they feed on the submerged form of *Subularia monticola* and *Crassula granvikii*. Although it has already been said that Hyrax feed on these two plants, since they occur both as a terrestrial and a submerged form, there results an interesting division of preference, each animal utilising almost exclusively one form only.

Onychognathus tenuirostris raymondi MEINERTZHAGEN
(Slender-billed Chestnut Winged Starling). From the forest up to 15,500 feet.

These birds are not permanently resident at high altitude. In the Teleki Valley a flock of a dozen birds fly up the valley to the Teleki Tarn, where they remain all day, and then at sunset they fly back to the forest. The same observations were made in the Mackinder Valley in January 1963, where the birds were seen to pass between the forest and the valley head every day for a week. No stomach contents were examined, but on several occasions the birds were seen feeding around the Teleki Tarn, where they collected molluscs either from *Lobelia* plants, or by turning over stones, where no doubt they also took some insects. Small heaps of smashed shells were found around the lake's edge, alongside flat stones, which are used by the birds to assist in breaking the molluscs open.

Algyroides alleni (Barbour)
(Alpine Meadow Lizard). 10,000 feet to 15,000 feet.

This is the only resident reptile in the Alpine zone and it is found amongst tussocks and under stones around the Tarns. An interesting colour variety exists amongst these reptiles, those around the Tarns being bright green, while those living in drier situations are the normal brown colour. The Museum of Comparative Zoology at Harvard has examined the specimens and found that the squamation of the brown and green forms is similar, and that lacking a complete altitudinal series, it is not possible to separate them (Edwards 1960, in correspondence). Stomach contents indicate that they feed largely on beetles and their larvae. Undoubtedly those living close to water feed to some extent on floating matter. These vertebrates are more dependent on the climate than the Mammals and Birds and are active only in sunny weather. Turning over stones while searching for this lizard in cold weather, the author found specimens in an almost completely moribund state.

To elucidate this problem further will require a more detailed study of population, which in turn must be related to the amount of available specific food. In this way it will be possible to calculate quantitatively the effect of the resident vertebrate and invertebrate populations on their habitats. Pearsall (1961) has written of this problem in the Masai Reserve in Kenya and Tanganyika, where he has shown that it is possible to calculate conversion ratios for cattle and the available vegetation; and although here it is concerned with a dry savannah habitat, many of the problems involved are similar to those found in the Alpine Zone, for due to low rainfall these lowland habitats are also unable to support more than a small number of animals. Payne and Ledger, working on game plains in Kenya, have found that, due to their migratory habits, the indigenous herbivores are able to utilise the vegetation cover better than introduced domestic stock. This is not surprising for, due to the large number of predators in the low savannah, cattle are herded and feed in close packs, thus denuding the habitat considerably. Wild game, on the other hand, move as they graze and spread out over large areas, and their selective habits result in the protection and preservation of the vegetable cover.

2. Herbivores and their Relation to Vegetation.

It has already been shown that the two commonest herbivores that exert appreciable pressure on the vegetation are the Mount Kenya Hyrax and the Groove-Toothed Rat. It is difficult to estimate the effect of the latter in relation to population size, since numbers are difficult to calculate on account of their mode of life. Around the shores of Lake Höhnel and Hall Tarns, where these animals are common, it is obvious that their effect is considerable. In July 1959 it was calculated that the runs of Otomys rats occupied and eliminated between 30% and 40% of the available ground for vegetable growth. The effect of denudation increased as the water was approached.

Although the Hyrax is a burrow dweller, it is possible to estimate approximate figures of population size. In making this calculation, any possibility of intermixing between communities is not considered, as in most situations colonies occupy ridge or moraine habitats, and by the nature of the intermediate terrain they are fairly effectively isolated from one another. A colony at Mackinder's Camp was observed between December 1957 and January 1958. PEARSALL (1961) gives details for the relation of nomadic pastoral Masai cattle-vegetation. The weight of human beings in a single manyatta lies between 1/40th and 1/50th of the weight of their cattle. Thus, the ratio of herbivore to vegetation will be a fraction of this, say 1/100th or less. In the moorland of Scotland the conversion ratio for sheep lies between 1/50th and 1/100th of the vegetation production. The same low figures might be expected for Hyrax on Mount Kenya.

Hyrax are very active in bright sunlight and in the early morning, when on most days the weather is bright and warm. Most of the animals in the colony emerge at this time and lie on the rocks sunning themselves. The following figures give the number of animals seen to emerge in the early morning at Mackinders' Camp (13,750 feet) at the head of the Teleki Valley, in January 1958:

1st day 78 animals
2nd day 87 animals
3rd day 72 animals
4th day 84 animals

(N.B. On dull days, with no direct sunlight on the colonies, very few animals appeared)

The average number of animals to emerge was 80, and assuming that most of the animals came out at some time over these four days, the figure of 80 can be considered as an average of the population size for this particular colony.

This group of animals forage between their colony and the stream below, and for a short distance up and down the valley, or on the rocks above their burrows. Except for the possibility of an occasional animal moving further afield, the area of a forage is approximately 500 yards square, i.e. 250,000 square yards. Thus an area of 250,000 square yards supports a colony of 80 animals. The average weight of animals is slightly above that for lowland species, being approxi-

Plate 21A. A young Mount Kenya Hyrax.

Plate 21B. A group of J.B. Sale's captive Mount Kenya Hurax Colony in Nairobi. The animal in the centre is threatening photographer showing us pointed 'tusks' and lower comb like incisors.

Plate 22. A-B. Female Scarlet Tufted Malachite Sunbird visiting her nest. The white material within the nest is the wooly tomentum stripped by the bird from the backs of *Senecio brassica* leaves.

Plate 22B.

Plate 23. Female Hill Chat.

Photograph by Dr. J. B. Foster

Plate 24. Mackinder's Owl.

mately 8 lbs. Assuming that one-third of the total number are young animals with an average weight of about 2 lbs, the total weight of herbivores supported by this area is 480 lbs.

Vegetation within five to ten yards of the burrow is eaten very low, except for those plants which are never eaten. Further away from the burrow the animals are very selective in their feeding habits, and although they go to the river to drink and to feed on water-side vegetation, a large strip of land between the burrows and the river is not used extensively in feeding. (The runs go direct from the colony to the water, with few side tracks). Thus, although the area of 250,000 square yards is the total forage area covered, a small percentage of this is actually visited during feeding (probably under 20%). This is due in part to the presence of *Otomys*, but more particularly to the large amount of coarse tussock grass and rosette plants that occur between the burrows and the water, which are unsuitable as food.

Stomach content of animals dissected were weighed (wet) and gave an average weight of .52 lb. (This does not take into account the small quantities taken by younger members of the community, or the fact that weighed wet, probably 20% to 25% of this weight derives from gastric juices). Since these animals feed mainly in the early morning and evening, twice this weight might be expected to be taken during the day. Thus, if 1.04 lb. of vegetable matter were consumed by each hyrax per day, the whole colony would eat up to 83.2 lbs. In the case of a slow, but steadily growing vegetation, unaffected by marked seasonal phenomena, a colony of animals grazing over an area of $\frac{250{,}000}{5}$ sq.yds., or 10.3 acres, would exert a considerable grazing pressure. This pressure may be roughly estimated as $\frac{83.2}{10.3} = 8.1$ lbs./acre/days. Expressed in terms of biomass, this area supports approximately $\frac{480}{10.3} = (46.6)$ about 50 lbs./acre within the region of Hyrax forage. Presuming the effect of other herbivores in this area to be negligible, this figure is very low. PEARSALL (1961) quotes the sheep uplands of Scotland as supporting about one animal per 2.2 acres. Biomass data for similar regions in East Africa is not, unfortunately, available, but some idea of how low these figures are can be obtained by examining the results of Game Counts conducted in the Royal Nairobi National Park from 1960 onwards, which have been carried out by a number of volunteers, including the author. SIMON (1962) includes as an appendix to his study of Kenya Wild Life a summary of the first year's figures. This gives a biomass on the whole Park of 73,000 lbs./sq.mile, or 114 lbs./acre, i.e. 2½ times greater than that found at 13,750 feet on Mount Kenya. However, it must be remembered that these latter figures have been derived from only a small area of the total forage zone of plains game, and represent a few months of equable conditions, with high game concentrations, followed by months when almost all the animals have moved south towards Tanganyika. It is still possible to speak of the Alpine Zone of Mount Kenya and the lowland savannah in terms of what PEARSALL (1961) has called "hard habitats". Both are greatly affected by climate, the former by drastic diurnal changes, and the latter by severe seasonal changes. The vertebrates in both habitats respond to these extremes in two ways, those in the former being restricted to a comparatively small area, conserve their forage by means of a low biomass, while those of the latter produce a similar effect by scattered grazing over a very large area. (Grant's Gazelle, *Gazella granti*, may range over an area of up to 100 miles).

It is expected that a similar situation will be found to exist if, at a later date, it is possible to form parallel estimates in the case of Otomys rats and Duiker, although, owing to the habits of the former and the shy nature of the latter, this may prove very difficult.

(*a*) *Population size and control.*

It has already been noted that the numbers of Hyrax and *Otomys* in their favoured localities is large, but that these numbers are spread over a large area, and that many of the mammal communities seem to be living effectively isolated from one another by virtue of the nature of the intermediate terrain.

Distribution of the fauna seems to be controlled by two main factors:

1. *The availability of food.* The food required by birds in the Alpine Zone is nowhere abundant, and the density of populations in any one area is directly controlled by the available food. The Scarlet Tufted Malachite Sunbirds are most abundant where insects are plentiful, and they fall off as the food supply drops. The same is true of the Streaky Seed Eater, which has a much less even distribution; this is controlled by the presence or absence of its main food plants, and since these are found chiefly in valley bottoms, the distribution of the birds is reduced accordingly. The Chestnut-winged Starling is also only found in areas adjacent to lakes and over water where molluscs are particularly abundant.

2. *The role of territory.* The sunbird is strongly territorial in its behaviour, and the Alpine Zone is divided into a large number of small territories (COE 1961). This means that these birds have a reasonably even distribution, except that the size of the territory

occupied by any one pair varies with altitude. Although such territorial behaviour is basically a breeding response in a region where food is restricted, the area demarcated must be closely connected with the amount of ground the bird needs to cover in order to acquire sufficient food to raise its brood. The abundance of food for these birds falls sharply with increase in altitude, and there is a corresponding increase in the size of the territory, which in turn leads to a lower density of birds over a unit area.

The number of any one species of bird seems to be maintained close to an "ideal" number, whereby the communities survive in an otherwise marginal region.

Since the number of animals in any one area is critical in respect of their survival, it is important to consider the factors that exert a controlling effect on the numbers of these communities. First and foremost amongst these is the obvious control pressure that predators must exert on Alpine resident communities. Creatures which fall within this group are:

Mammals.

Leopard. (Felis pardus pardus L.). A single pair seem to live in most valleys. From examination of their lairs they seem to be resident in the Alpine zone, and range from the moorland up the main peaks (16,000 feet). Examination of their faeces reveals that their main food consists of Hyrax and Otomys rats. They almost certainly feed also on Duiker. (In July 1959 the remains of a half-consumed Duiker were found under a ledge in the Gorges Valley.)

Wild Dog. (*Lycaon pictus lupinus* THOMAS). Several packs live in the Alpine Zone although they may not be permanent residents. In December 1957 a pack of eight dogs roamed in the area of the I.G.Y. Expedition Camp II, at an altitude of 12,750 feet in the Northern Naro Moru Valley, and on one occasion when they came unpleasantly close to the camp the Transport Officer had to drive them off with a revolver. Since they follow the pack technique of hunting, these mammals almost certainly feed largely on Duiker and, to a less extent, on Hyrax when they catch them in the open. On the Northern slopes where the moorland reaches 14,000 feet they feed on visiting herds of Zebra, Eland and smaller gazelles.

Lion. (*Felis leo* sub-sp.?). Lion spoor have been seen in the snow above Top Hut at 16,000 feet (BAKER, personal communication 1961), but it is doubtful whether this represents any more than an occasional straggler. GANDER DOWER (1935) mounted an expedition to look for the "Lion of the Bamboo", a creature said to be of the build of a lion, but with the spots of a leopard. Although the expedition did not collect any positive evidence of the existence of this creature, there is undoubtedly a lion living in the upper limits of the Montane forest that may venture into and feed in the Alpine zone. In July 1959 the author met two honey hunters on the Chogoria track, below the Gorges Valley, who claimed that they had seen a "Simba ya msituni" (Lion of the forest) at about 10,000 feet.

Red River Mongoose (*Atilax paludinosus rubescens* HOLLISTER).

Not previously recorded on the mountain, but in January 1958 the author saw these animals at the foot of the Gorges Valley (11,750 feet). They probably prey on Otomys rats on the side of streams and lakes. Since there are no specimens in existence from this region, it seems that this record must represent a very small population indeed.

Birds.

Mackinder's Owl. (*Bubo capensis mackinderi* SHARPE) (Plate 24)

Resident in the Alpine Zone: at least two pairs in the upper Teleki Valley. Examination of their pellets indicates that they feed almost exclusively on Otomys rats. Birds have also been heard at night in the Nairobi, Gorges and Mackinder Valleys.

Augur Buzzard. (*Buteo rufofuscus augur* RUPP.)

These birds usually visit the Alpine zone in the day, but in January 1963 the author observed a pair of birds at the head of the Mackinder Valley that remained resident for the whole period of the author's stay (5 days). During this period the bodies of three freshly killed Hyrax were found in the area. Since the population at the time was not large, it would seem likely that these birds are one of the most important predators of Hyrax on the mountain. In addition they also take Otomys rats, and probably an occasional bird.

Verreaux Eagle. (*Aquila verreauxii* LESSON)

Not common. One pair was seen breeding on the Hall Tarns cliffs at 13,700 feet in January 1958. They feed largely on Hyrax.

Lemmergeyer. (*Gypaetus barbatus* LINNAEUS).

Resident in small numbers in the Alpine zone. Those that are present probably feed largely on Hyrax.

Although individually these predators are very efficient hunters and must account for large numbers of Hyrax and *Otomys*, since the total number of all predators in the Alpine zone cannot be more than 40—50 individuals, their gross effect as a means of population control cannot be great.

Disease

Disease may be a secondary factor in such control, but birds and mammals dissected on the mountain were remarkably free of endoparasites (whereas Hyrax collected in lowland areas are usually infested with Cestodes). Negative evidence of this sort cannot, obviously, be taken as proof that epidemics do not exert a controlling effect on populations, though the comparative isolation of individual communities, the absence of arthropods that may act as intermediate hosts of parasites, and the marked temperature changes to which faeces are submitted are all features that would seem to rule out disease as a factor of any great significance in population control.

(*b*) *Breeding as a control factor.*

If neither the effect of predators, nor disease can be considered to be factors of primary importance in maintaining population numbers, then it would appear that such control must be exerted within the community itself. During field work on Mount Kenya the reproduction rates of both the Hyrax and the Scarlet Tufted Malachite Sunbird have been investigated, and this data seems to suggest that for these species the number of young has been reduced below the modal number for the genus, thus controlling the size of the population to near the ideal number for these animals.

LACK (1959) has considered in great detail the factors controlling the number of offspring by birds and mammals. Having eliminated previous ideas of physiological control, or longevity factors, the available evidence clearly indicates that clutch size has evolved as the number of young that the parents can reasonably hope to support. It has also been found that when a large sample of clutch sizes was observed, in cases where the parents endeavoured to raise more than the minimum clutch, fewer rather than more of the young tended to survive. This is a reasonable supposition, for each parent will have a maximum food collecting capacity and thus in larger clutches each bird will tend to get less food. This must be a factor of great significance in a hard or marginal habitat like the Alpine Zone of Mount Kenya, where the available food for a given bird in a unit area is restricted. This factor will require the bird to forage for greater distances, and hence the amount of food it can provide for its brood will also be restricted. MOREAU (1947), KLUIJVER (1933), GIBB (1950) and KENDEIGH (1952) all confirmed this fact when they related the number of feeding visits to increase in brood size. They found that the increased number of feeding visits did not rise in proportion to the increase in size of brood, and that each young bird tended to get less food and hence their chances of survival decreased.

Although LACK found that some Passerine birds have a variable clutch size, depending upon the seasonal availability of food, this is not a factor of great importance in considering the Mount Kenya communities for, at any rate over the author's period of study, the climate of the Alpine Zone seems to be remarkably constant, thus producing a more or less constant (even though at a low level) food supply.

In many cases it appears that the length of day is an important factor influencing, through the time available for the collection of food, the size of clutch; and in situations where two broods are produced annually, such variation may have a profound effect. Although Mount Kenya is on the Equator and hence the fauna has a constant 12 hours of daylight throughout the year, the diurnal temperature variation is considerable and reduces the time available for food collection to an appreciable extent. On fine days the sun does not shine into many valleys until quite late in the morning, and then sets as much as two hours before dusk, leaving as little as eight hours for food collection.

In the case of fairly large birds, whose modal clutch number is low, this can often be related to the type of food required for the young. SCHMAUS (1938) found that Buzzards usually only raise one out of two eggs, while in good seasons up to three eggs are produced, all of which survive. The same is true of Storks (SCHUZ 1942) and Swifts (LACK 1951).

Similar studies have been made on Mammals and the factors controlling litter sizes seem to be the same as those affecting birds. BAESTRUP (1941) found that the number of young produced by the Arctic Fox in Greenland varies with the frequency of Lemmings in the season concerned. Similar observations were made by STEVENSON-HAMILTON (1937) in the Kruger Park on Lion, which produced more young in seasons when game was abundant. To a large extent the author's own observations support the opinions outlined above on population control.

In the Alpine Zone the Scarlet Tufted Malachite Sunbird is the only resident member of the Nectariniidae. Nests are built in Erica bushes in the moorland zone, and at higher levels they are constructed in the tops of grass tussocks, or amongst the dead leaves below the terminal rosettes of *Senecio keniodendron*. Since these birds exhibit distinct territorial behaviour throughout their range, finding their nests does not present great difficulty. Birds have been recorded nesting from 11,000 feet to 14,000 feet.

During the course of field work the author examined the following nests, which he found to contain either eggs or fledglings:

Date	Locality	Site	Content
Dec. 1957	Northern Naro Moru Valley	Grass tussock	1 egg
Dec. 1957	Northern Naro Moru Valley	Small Erica bush	1 egg
Dec. 1957	Teleki Valley	Grass tussock	1 Fledgling
Dec. 1957	Teleki Valley	Senecio	1 egg
Jan. 1958	Hall Tarns	Grass tussock	1 egg
Jan. 1958	Hall Tarns	Grass tussock	1 Fledgling

In addition to these records with actual eggs or fledglings in the nest, in all the localities cited above parents have been observed in company with a single young bird. This suggests that these birds regularly produce only one egg; and this may be explained as a response to the availability of food in the region of the nest. These birds obtain most of their food from either *Lobelia* or *Senecio* inflorescences, and when the small Lycaenid butterflies, *Harpendireus aequatorialis*, are abundant they feed their young almost exclusively on these insects. The amount of food they need to collect for their young is considerable, and in an environment where insects are not in plentiful supply this reduction in the number of their offspring would confer an advantage on the young bird and increase its chances of survival, while at the same time effectively controlling the population of these birds in the Alpine Zone.

Amongst lowland Nectariniidae, although many do produce two or three eggs, one is not an uncommon clutch size amongst other species which do not live in the apparently hard habitat of Mount Kenya species. However, reduction in the size of clutch may not only be a factor controlling the population in an area where food is short, but it would work equally well in an environment where intra-specific competition for food and territory is intense.

The study of Hyrax provides more evidence that the size of the litter has been reduced in response to competition for available food and in the interest of maintaining the size of the community at a level that will not denude the habitat. It is only possible to estimate the number of young produced by the dissection of Hyrax, or with rather less certainty by observation of the colonies. Amongst animals shot and dissected in the Alpine zone the following information on breeding was obtained:

Date	Locality	State of Breeding
Dec. 1957	Northern Naro Moru Valley	2 Females with 1 embryo each.
Jan. 1958	Teleki Valley (13,750 feet)	1 Female with 1 embryo
July 1959	Gorges Valley (14,100 feet)	1 Female with 1 embryo

In addition to these positive records obtained from pregnant females, observation of colonies also suggests that the commonest number of young produced is one. In the early morning and the evening, especially when the sun first strikes the colonies, small family parties come out on to ledges to sun themselves. These groups usually consist of a male accompanying several females and their young. Although the offspring run up and down the ledges, they spend a great deal of time suckling their mothers and except for the odd occurrence of an occasional female with twins, the proportion of young appears to be one to each female. Also on occasions when the females go down to streams and lakes with their young (e.g. in the Teleki Valley, 13,750 feet and the Gorges Valley 13,000 ft.) it is possible to observe the offspring ratios, and here again, the presence of a female accompanied by more than one young animal was seen to be rare.

More recently SALE (1965) has described the breeding behaviour and activity of captive specimens of the Mount Kenya Hyrax collected in the Mackinder Valley. Three females caught on the mountain produced two babies each in captivity in Nairobi. Since their capture in January 1963 they have produced from 3—1 young each. The subject cannot, however, be closed here for the Hyrax in the Mackinder Valley are sparse compared with the large and dense colonies of the Teleki and Gorges Valley. It would indeed be interesting to show that the comparative isolation of suitable rocky habitats for Hyrax leads not only to a varying population density, but also a local variation in the number of offspring.

ASDELL (1946) in discussion of the breeding of the South African Rock Hyrax gives a mean of 2.44, with a mode of 2 and a range from 1—6. Dr. John SALE, of the University College, Nairobi, has found in his studies of Hyrax that for lowland species the modal number in Kenya is also 2.

We may conclude from this information that the Mount Kenya Hyrax has developed this low number of offspring in response to pressure from its harsh environment. Such a reduction in the number of young would, in a region of slow growing vegetation, confer a considerable survival advantage on the young. In addition, since it appears that predation and disease have but minimal influence on the population size, such a reduction would be of profound importance in maintaining the number of animals close to the "ideal" for this habitat.

As yet the fauna of Mount Kenya is very poorly known, and any account of basic factors in the Alpine zone must be necessarily preliminary in its scope. Even the smaller vertebrates are little known, and the moorland which seems to have a particularly rich rodent fauna, will no doubt yield rich results when it is systematically collected. In fact, apart from SALT's paper on Kilimanjaro (1954), the Alpine regions of the East African mountains remain an almost virgin field of the zoologist and, more particularly, the ecologist.

DISCUSSION

The peak materials of Mount Kenya have recently been dated in the United States by EVERNDEN and CURTIS, who have produced a figure of 2.7 million years (BAKER 1963 — Personal communication). There is evidence to suggest that at this time the climate in East Africa was drier and warmer than at present. Since, however, Mount Kenya must have been at least 23,000 feet in height before its later intense glacial erosion, its upper slopes must have still been very cold, even although they were not glaciated. Considerably later, and contemporaneous with the European glaciations, an ice sheet covered this and other East African mountains.

It now seems to be generally agreed that the major glacial phases were of comparatively recent occurrence, and that as late as 20,000 years ago the last glaciation was still in full force (MOREAU 1963). This means that the present mosaic of isolated Afro-alpine and Afro-montane plant and animal communities were still to some degree approximated, even if not in direct connection. Recently LIVINGSTONE (1962), using cores taken from Mahoma Lake that lies at 9 840 feet in the Ruwenzori range (of considerably greater age than Mount Kenya), has shown that this region became deglaciated approximately 15,000 years ago. If we accept that general ice retreat began at this time and, except for periodic halt phases or minor readvances, the present Alpine Zone of Mount Kenya has also been progressively colonised. If the Alpine plant communities have developed in this short space of time, it is difficult to reconcile this suggestion with the high percentage of endemism that occurs on this and on other mountains. Under the rigid Alpine climate the rate of selection would undoubtedly have been greatly accelerated, even over such a short time as 15,000 years. So far pollen analytical studies have only shown a general movement downwards of the montane and moorland biomes, though if these habitats were depressed, the alpine region was also, no doubt, depressed at least to some degree at the same time.

HEDBERG (1961) has examined endemism on the East African mountains and it is interesting to note the existence in their Alpine flora of 23 endemic species on Elgon, 7 on the Aberdares, 13 on Mount Kenya, 13 on Kilimanjaro and 3 on Meru. The higher percentage of endemism on Mount Elgon may well indicate that the route by which plants from the North migrated from the Ethiopean highlands, via the Immatongs and intermediate ranges, led first to this mountain and later radiated from hence to other mountains East of the Rift in East Africa. In addition it should be noted that the higher mountains bear a higher percentage of endemic species. Since the higher the mountain, the more extreme the climate, this may again suggest that climate has played an important part in the apparently rapid speciation in these regions. Although Mount Kenya and the Aberdares are only separated by a ridge of land 50 miles wide, that never falls below 5,000 feet, the former bears almost twice as many endemic species as does as latter.

It appears also that during the glacial period many plants that were possibly new to the African flora found their way to the East African highlands and were later isolated due to the return of a more equable climate. This fact is exemplified by HEDBERG (1961) who notes that the Afro-alpine phanerogamic flora is of extremely complex origin, deriving from the following elements: 20% Afro-alpine, 13% Afro-montane, 6% South African, 4% Cape element, 3% South Hemisphere temperate element, 7% Mediterranean element, 15% North Hemisphere element, 2% Himalayan element, and a cosmopolitan element of 30%. Thus when we consider the flora and indeed, perhaps to a lesser degree, the fauna of Mount Kenya, the complexity of their derivation and the apparently rapid changes in speciation that have taken place since their comparatively recent isolation must continually be remembered as an undertone to all studies of the Afro-alpine biota.

MOREAU (1963a, 1963b) has examined the distribution of the African avifauna giving special attention to Equatorial Highland regions. He points out that the lowland rain forests surrounding Ruwenzori and Cameroon have provided a refuge for montane forest species. The more Easterly mountains which have come under the influence of a drier lowland climate have been more effectively isolated by large areas of intervening savannah. The power of isolation of these intermediate lowland regions is illustrated by reference to Usambara in Northern Tanganyika where 35 species of passerine birds live in the montane-highland forest boundary at 4,500 feet. Of these species 26 descend to 3000 feet, while only six occur at 2000 feet, suggesting that the lower limit of their downward dispersal is about 2,500 feet.

In spite of the apparent barriers, widely separated land masses are found to harbour similar subspecies, and in some cases even species that cannot be accurately distinguished. The Scarlet Tufted Malachite Sunbird (*Nectarinia j. johnstoni* SHELLEY) of Mount Kenya cannot be differentiated from that in either the alpine regions of the Aberdares or Kilimanjaro (WILLIAMS 1951). Other subspecies are recognised from Ruwenzori and Kivu (*N.j. dartmouthi* GRANT), and South Western Tanganyika (*N.j. salvadorii* SHELLEY).

MOREAU (1963a) accounts for the present distribution of the Afro-Montane avifauna by postulating an extension of montane forest throughout East Africa, from Northern Kenya to Nyasaland,

and the Cameroons and S. W. Angola. A lowland forest connection also appears to have existed between the Congo and the East African coast via the Lake Victoria Basin.

Both the above author and CARCASSON (1964) are in agreement in suggesting the persistent isolation of the Ethiopean highlands from those of East Africa, the former from their bird distribution and the latter from the point of view of their butterflies.

COE (1966) has briefly considered the distribution of mammals in the East African mountains, and has pointed out that the Crested Rat (*Lophiomys imhausi* M-EDW.) displays a distinct zoogeographical affinity between the Kenya Highlands and those of Ethiopia and Somalia. The distribution of Western and South Western mammals shows a marked similarity to that described for the avifauna.

In the past the distribution of the giant *Senecio* species *keniodendron* and *brassica* has been used as an indication in the recognition of an upper and a lower Alpine Zone. While in general terms this may be a convenient means of subdividing this extensive region, it must be remembered that the distribution of these plants is controlled by the presence or absence of their most favoured terrain, rather than purely by factors of altitude. This is well demonstrated in the case of *Senecio brassica*, which favours flat or slightly sloping waterlogged ground, while *S. keniodendron* is found only on well-drained sloping ground, with a good supply of surface water. Only where these two situations border one another do these two species occur close together. The fact that *S. brassica* is not usually found above 12,629 feet (3,850 m) is due entirely to the steep slope of ground over 13,500 feet which completely eliminates the water-logged surfaces required by this plant.

The same feature can also be demonstrated with regard to the distribution of *Lobelia keniensis* and *L. telekii*, the former no longer occurring over an altitude of about 13,500 feet where suitable areas of wet boggy ground are not generally found. While it is true that stream banks are suitable situations for colonisation by these plants, the very low temperature of melt water from the glaciers, and the cold valley winds are probably additional factors of importance in eliminating these plants from associations at high altitude.

Apart from features controlling the distribution of the individual species described above, the same is true of vegetational associations whose distribution is directly related to the mountain's topography and climate. For instance, plants associated with glacio-fluvial surfaces are not found below about 13,500 feet, though this is not necessarily an altitudinal feature, but rather one related to the distribution of the substrate on which they are found. In this particular instance glacio-fluvial fans do not occur below this altitude at the present time, though they may well have done so in earlier times. Hence, the plants associated with them have now become limited to a distinct altitudinal zone. The tussock grass community that is so characteristic of rocky ridge tops extends from the lower Alpine well into the upper Alpine zone. Here again, the phenomenon is not one that can be related to altitude, but rather, in this case, to the nature of the surface upon which the community occurs, and to the climate. If the ground of these rocky regions is examined, it is noticed that with increase in altitude, frost heaving becomes appreciably more common until at over 14,000 feet the ground surface is almost 100% disturbed. Without a doubt the more recent glacial history of these higher regions also plays a part, for it has provided the predominantly fine fraction that is so susceptible to frost phenomena.

The term "zone" which is so often used in vegetational descriptions of equatorial mountains is also somewhat ambiguous since it suggests continuity (BOUGHEY 1955). Basically the term "zone" is convenient in such classifications provided that it is noted that in the Alpine Zone such a region is made up of a mosaic of very distinct plant communities. The importance of the dissection of the land surface in the Alpine Zone is perhaps best illustrated when this region on Mount Kenya is compared with similar altitudes on Kilimanjaro. The latter consists largely of the recent crater, Kibo, which is little dissected and preserves an almost perfect cone-like form. For this reason the vegetation in what SALT (1954) has called Alpine desert is remarkably constant.

The climate of the Alpine Zone of Mount Kenya is essentially one in which marked seasonal phenomena are absent, but in which extraordinary diurnal changes of temperature occur. In conjunction with these temperature fluctuations the humidity also rises and falls in approximately the same cycle. It has already been shown that the difference in ground to air temperature is also a vital factor in deciding the type of plant community that will survive in a given area. The deeply dissected surface of Mount Kenya provides a remarkably large number of niches, and apart from the influence of a regular diurnal climate, and the marked difference in ground to air temperatures, the climate near the ground in situations quite close together may be very different. HEDBERG (1963) made temperature recordings in the Teleki Valley (13,776 feet) at two stations placed 250 metres apart, one on the valley floor and one on the valley wall. The resulting data shows that the temperature in the valley bottom had an amplitudinal variation twice that recorded on the valley wall. The author has made similar observations in the Mackinder Valley, where thick hoar frost was frequently recorded in the valley bottom, with a ground temperature as low as —9 °C (15.8 °F), while 200 feet away at the base of the valley wall no frost was visible and the temperature did not fall below —1 °C or

—2 °C. The remarkable difference in ground climate was also clearly reflected in the degree of needle ice formation which was severe on the former site, decreasing in severity until it could not be seen at all on the latter.

Such important micro-climatic factors explain in large part the complicated mosaic of communities that occur in the Alpine Zone, and it is for this reason that the plant communities have been treated more from a topographical point of view than from one of altitude. Provided the microclimate of protected niches at 12,000 feet and 15,000 feet are similar, they will support similar communities almost irrespective of their altitude.

However, while the diurnal climate is very marked, the climate over an average year remains remarkably constant from day to day. The fact is well illustrated if flowering records (HEDBERG 1957; COE, personal field records) are examined, when it is found that of 180 species occurring in the Alpine Zone, 85 species have been recorded as being in flower from ten to twelve months of the year (i.e. 47%). This percentage might well be higher were it not for the fact that there are still serious gaps in our knowledge of what species are in flower during months when access to the mountain is difficult. In addition, a certain number of rare and little collected plants are included in the total number of species occurring, and if these are eliminated, the percentage of plants found to be in flower during ten to twelve months of the year is increased to nearly 80%.

A similar feature appears to be exhibited by the Scarlet-tufted Malachite Sunbird, which seems to produce eggs at any time of the year. Certainly in an environment with a diurnal rather than a seasonal climate, there will be little seasonal variation on available food, which is usually the factor of highest importance in determining when young of either mammals or of birds will be produced.

While the presence of megaphytic plants is a very obvious feature of the Alpine Zone, by far the greater percentage of flowering plant species are of a rosette or creeping form of growth. The common occurrence of this habit in Alpine regions is well known, but its cause does not appear to have undergone exhaustive investigations. Recordings of ground to air temperatures in the Mackinder Valley have shown that plants close to the ground are subjected to rapid changes of temperature which may not be just a diurnal change, for when the weather is inclement the plant may be submitted to a temperature change of as much as 20 °C several times in one day. While this would appear to be a disadvantage, the insolation at ground level during periods of unobscured sunlight would seem to confer an advantage on the plant with a rosette habit. Plants brought to Nairobi were maintained in two groups, one of which was kept in constant room temperatures, while the other was kept at room temperature during the day and cooled to freezing at night. It was found that the latter plants maintained their rosette habit, whereas the former group elongated rapidly. While such observations may not be taken as conclusive evidence that diurnal temperature changes are the sole factor in creating the higher percentage of the rosette growth forms found in the Alpine Zone, it certainly appears to be a factor of primary importance.

Even plants with a megaphytic growth form show marked adaptation to diurnal temperature phenomena. The rosettes of both Giant Senecio and Giant Lobelia all close at night as the temperature falls, and the author has noted that the degree of their closure is directly related to the temperature. HEDBERG (1963) has observed that at dawn, as soon as the sun falls on a *Lobelia* rosette, the leaves thus illuminated immediately begin to open even though the air temperature is still below freezing. The rapid insolation at the surface of the leaf is no doubt as marked as that observed at ground level. The same is true of even the small herbaceous *Senecio* and *Helichrysum* species, whose leaves are tightly folded over the growing points at night; and it has already been noted that there is a distinct altitudinal tendency for the degree of hairyness of these plants to increase with altitude.

It also seems possible that the water that collects in the rosettes of *Lobelia keniensis* may well protect the basal parts of the leaves, for at night the water freezes above and to the outside, leaving the inner and basal surfaces protected by a layer of unfrozen water.

The wide variations in pigmentation and stature that occur amongst the Alpine flora with changes of altitude may well also be a response to temperature. It may too be found that ultra-violet and infra-red radiation also play an important role. It has been shown under laboratory conditions that plants submitted to ultra-violet radiation become stunted and the leaves become hairy (v.d. VEEN & MEYER 1959), and it may well be shown that the same process is taking place on Mount Kenya. Unfortunately such observations are costly in operation, and up to the present it has not been possible to arrange for such measurements to be carried out on Mount Kenya. An intensive study of the bio-climatological phenomena on the equatorial mountains could well prove to be a study of primary importance in our understanding of the Alpine biota.

The extreme climate of the Alpine Zone has also had a profound effect in controlling soil formation and distribution. In fact, we may say that glacial action has been the primary soil-forming factor, and that the subsequent rate of soil formation and distribution has been brought about by the agencies of water, marked diurnal temperature variation, and the deeply dissected glacial topography. The plant communities that have become established in the Alpine Zone have done so on the surfaces created by these soil-distributing agencies. It is a very obvious fact in this region that the establishment of vegetation is a marginal business, and that whenever the climatic extremes cause extensive frost-soil phenomena, or where the exposure to these extremes is great, plants do not become established. This is essentially a matter of altitude, for with an increase in altitude, the degree of exposure and extremes of climate increase roughly in proportion.

While it is difficult to follow the phases of vegetation establishment and development at lower altitudes, it is possible to see the primary stages of plant colonisation on the newly exposed moraines of the peak region. Since the mountain has in comparatively recent time been exposed to intensive glacial action down to an altitude of approximately 10,000 feet, the phases of colonisation taking place at the present time close to the glaciers must be similar to those that took place in the wake of the retreating ice at the glacial maximum. Of particular interest in this region is the approximation of the newly colonised recent moraine and the rocks of a much greater age, now densely covered with lichen, which strongly suggest a recent glacial advance. It will be of great interest to carry out similar observations on other equatorial mountains where ice still exists, to see if similar vegetational evidence of a glacial advance can be found.

The invertebrate and vertebrate fauna of Mount Kenya are still very poorly known. Even the small mammals which have been generally well collected in Kenya, have been little studied. Perhaps our most complete faunal knowledge is of the avi-fauna, through the energy and enthusiasm of such men as JACKSON, MEINERTZHAGEN and CHAPIN, although even here the zoogeographical relations of the avi-fauna of adjacent Equatorial Alpine regions has, except for the work of MOREAU, been only superficially documented. It is hoped shortly to start a programme of work that will involve comparative studies of the vertebrate fauna of all Equatorial mountains East of Mount Elgon. This study, when completed, should be of great zoogeographical and evolutionary interest.

However, while but scant attention has been given to studies of vertebrate fauna, the invertebrates are to all intents and purposes unknown. This is well illustrated by the fact that in December 1957 the author collected specimens of a small bivalve mollusc which occurs in large numbers in Alpine lake sediments. This mollusc was described by KUIPER in 1960 as *Pisidium artifex* KUIPER. In addition, several species of *Vicariihelix*, one species of *Eucobresia* and one species of *Limicolariopsis* await further examination and the possible raising of new species.

Apart from the small *Pisidium artifex* KUIPER that occurs in the Alpine Lakes on Mount Kenya, there are also a large number of interesting Crustacea, many of which are to be found as high as 15,000 feet. Heinz LOEFFLER, from Vienna University, visited most of these waters in consultation with the author in 1961, and in subsequent correspondence he has stated that many new species await description. Even the highest main lake on the mountain, Hut Tarn (14,750 feet) has yielded nine species. Amongst the species recovered by LOEFFLER and the author, the genera *Candanopsis* and *Maraenobiotus* (Ostracods and Harpacticoids respectively) show distinct species on Mount Kenya, as they do also in the palaearctic region and the Andes.

The same can be said to be true of the diatom flora of these lakes. CHOLNOKY (1960), from a sample collected at the Hall Tarns (14,300 feet), described over fifty species, of which he recognised five new species. Here the affinities appear to be Northern and Himalayan.

Since the identification and description of the invertebrates collected on the mountain are as yet far from complete, this work has been largely confined to a consideration of the role played by vertebrates in the Alpine region. These creatures can be considered as falling into three categories: 1. Resident vertebrates (9 species); 2. Regular visitors, or resident for short periods (8 species); 3. Occasional stragglers (5 species). It is only those vertebrates which fall into the first category that play any significant role in the ecology of the zone. Just as the severe climate plays an important part in limiting the growth rate and growth form of the alpine vegetation, so in turn this same restriction is placed upon resident vertebrates. This restriction is reflected in the marked spectrum of habitat and food preference exhibited by the Alpine mammals and birds. Undoubtedly, it is by this phenomenon, well known in other extreme habitats, that inter-specific competition is virtually eliminated and the habitat is preserved, carrying a small number of species but a large number of individuals.

The success of the Alpine vertebrates in their habitat is well illustrated by the fact that there is little mingling of this community with that of the moorland community where these two very distinct zones meet. Perhaps the most interesting separation is demonstrated by the two species of Hyrax which occur on Mount Kenya. The arboreal animal, *Dendrohyrax arboreus crawshayi*, is found in the montane forest but does not emerge on to the moorland, while the Rock Hyrax, *Procavia johnstoni mackinderi* THOMAS, only occurs at the upper border of the moorland zone at about 12,000 feet. The altitude restriction of the Rock Hyrax is due to the fact that suitable sites for occupation only exist within the limits of major moraines and lava cliffs. Thus the two animals are effectively separated by 1,000 feet of moorland (COE 1961). Undoubtedly the Rock Hyrax migrated to its present position some time after the main glaciers retreated through the Northern forest gap. By contrast, on Ruwenzori where no forest gap exists, the Tree Hyrax, *Den-*

drohyrax arboreus ruwenzori NEWMAN, has left its arboreal niche and occupied the boulder habitat of the Rock Hyrax.

In October 1963 the author visited Mount Kenya with Tom HARRISON, from the Sarawak Museum, and his immediate comment on seeing the moorland zone was that it so closely resembled the terrain on Mount Kinabalu (13,450 feet) in Borneo. In addition, the niches occupied by both mammals and birds were paralleled in almost every case. The same appears to be true of Himalaya where SWAN (1961) noted with some care the niches occupied by invertebrates, and in turn the relation of vertebrates to these food sources. In comparing the Mount Kenya vertebrates, it is found that in almost every case they have a counterpart on the Himalayas. It is hoped that shortly a programme of work will be started when comparative studies will be made of the vertebrate fauna of the East African high mountains and those of Borneo. The study of Afro-Alpine fauna in all its varied aspects is still in its infancy, and it will be many years before a clear comparative picture can be drawn.

SUMMARY AND CONCLUSIONS

1. The work is introduced by considering the discovery of Mount Kenya and subsequent exploration by travellers and scientists up to the present time.

2. The introduction is followed by a brief summary of the physiography, geology and glacial geology of the mountain.

3. Against the background of the early geological and later glacial history of Mount Kenya, the vegetational zones and communieties to be found on its slopes are described.

4. The classification of HEDBERG (1951) is maintained in the definition of a separate moorland zone (Ericaceous), a lower Alpine and an upper Alpine zone. It is shown that the recognition of the lower and upper Alpine zones, based upon the distribution of Megaphytic *Senecio*, is related more to the topography of the mountain than to altitudinal phenomena.

5. Since, due to vegetational creep in protected situations, many plants from the Ericaceous zone have entered into the lower Alpine zone. The description of the Alpine vegetation is prefaced by a description of moorland communities.

6. The most obvious plants in the Alpine zone are species of *Senecio, Lobelia, Alchemilla* and *Helichrysum*. The main features of these plants are briefly described.

7. It appears that certain characteristic morphological adaptations to high altitude are exhibited by the vegetation on Mount Kenya. These are discussed with regard to the Alpine zone of this mountain, and are briefly compared with SALT's observations on Kilimanjaro and related to the various surfaces on which they occur. These include valley walls; valley floors; glacio-fluvial areas at valley heads; solifluction terraces and other damp, almost flat ground adjacent to streams; ridge tops; and lakes and tarns.

8. The description of the main vegetation types is concluded with a description of the small and restricted Nival Zone. This zone is considered to start at the point where signs of recent glacial advance can readily be distinguished (TROLL 1958).

9. The climate of the Alpine Zone of Mount Kenya is essentially diurnal, with heavy insolation by day and frost by night. Seasonal phenomena are virtually absent.

10. It is shown that current research, particularly in the field of Palynology, suggests that 10,000 to 11,000 years ago the tree line on the East African mountain was from 1,650 to 1,968 feet lower than it is today. Such a lowering would indicate a drop in temperature of up to 4°C at least.

11. Temperatures recorded during the I.G.Y. Mount Kenya Expedition at three stations from 10,000 feet to 15,650 feet are

enumerated. A slight deviation from the International Standard Atmosphere calculations at the highest station is probably explained by cold glacier winds affecting this position.

12. Air temperatures recorded do not adequately reflect the Alpine temperature regime. The temperature near the ground shows great variation in amplitude, owing to insolation. This phenomenon, which has been noted on many other high mountains (GIEGER 1957), is also demonstrated in respect of Mount Kenya.

13. Rainfall data for Mount Kenya is now very good, thanks to a series of rain gauges maintained by the Ministry of Works. Figures obtained on the Naro Moru and Sirimon tracks show a correlation with lowland rainy seasons. There would seem to be considerable variation in the amplitude of wet seasons.

14. Isohyet diagrams show remarkable agreement with the main vegetation zones. To the South-east, where the montane forest is thickest, the highest rainfall is recorded. This varies from 90″ in an average year to 150″ in the unusually heavy rains of 1961. While such variation is to be expected, the main Isohyet pattern remains remarkably constant. High rainfall in the East accounts for the extensive moorland zone found in areas with this aspect, just as the low rainfall to the North occurs where there is a gap in the montane forest. The effect of mist and other "occult" precipitation, and of wind, is also briefly discussed.

15. The relation of climate to Alpine vegetation is discussed. It is pointed out that Equatorial mountains throughout the world which possess a diurnal climate show a tendency to produce megaphytic forms and a large number of plants exhibiting a rosette habit. The occurrence of the rosette habit seems to be closely related to an inhibiting effect on the elongation of internodes produced by diurnal temperature changes. When this diurnal change is removed the plants all elongate rapidly.

16. The megaphytic habit so common in the Alpine Zone seems to have been derived from many branched forms that occur in the montane forest. The two distinct rosette and erect habits are two distinct responses to the Alpine climate which are equally effective in protecting the plant from these extremes.

17. The Alpine soils of Mount Kenya are all largely derived from glacial activity. In the vicinity of present glaciers it is still possible to see the large quantities of glacial flour that are derived from this source. Valley heads show large glacio-fluvial deposits produced by halt stages in comparatively recent time.

18. Just as the diurnal climate has a marked effect on the vegetation, it is also the most important soil forming climatic factor at the present time. Alternate freeze and thaw causes extensive soil creep on slopes which, together with surface water, acts as an important soil transporting agency. Soil polygon and needle ice

formation play an important role in soil disturbance, which in exposed situations in the upper Alpine zone prevent the establishment of vegetation and lead to intense erosion of ridge tops.

19. The structure and chemistry of Alpine soils are briefly outlined.

20. It is shown that at the foot of the Tyndall and Lewis glaciers the primary stages of plant colonisation can be followed. The evidence derived from this study points to a recent readvance of glaciers in the peak region.

21. Other important phases of the establishment of vegetation in the Alpine zone are discussed. It is pointed out that while it is not possible to see the complete picture of colonisation, that derived from the study of old moraines, and mobile surfaces, assists in an understanding of this phenomenon.

22. The effect of the Alpine climate on the invertebrates and vertebrates is discussed. Due to incomplete knowledge of the former, attention is focussed chiefly upon the vertebrates.

23. Vertebrates in the Alpine zone, while few in species, appear to be large in the number of individuals. A remarkable spectrum of habitat and food preference preserves the habitat and virtually eliminates wide specific competition for nutriments.

24. The main Alpine vertebrates are described and their relation to predators in the zone are discussed. The control of numbers seems to be effected through a reduction in the number of offspring.

25. Preliminary observations suggest that where Hyrax colonies occur, a biomass of 46.5 lbs./acre can be expected.

BIBLIOGRAPHY

AFZELIUS, K., 1925. Einige neue Senecionen von Kenia und von Mt. Aberdare. *Svensk bot. Tidskr.* **19**: *419—422.* Uppsala.

ALLEN, G. M., 1939. A Check List of African Mammals. *Bull. Mus. Comp. Zool. Harv.*, **83.**

ALLUAUD, Ch. & JEANNEL, R., 1915. Le Mont Kenya en Afrique Orientale Anglaise. *Rev. Gén. Sci.*, **25**: *639—644.* Paris.

ARNELL, S., 1956. Hepaticae collected by O. Hedberg et al. on the East African Mountains. *Ark. f. Bot.* ,Serie 2. **3**, nr. 16.

AUTHUR, J. W., 1921. Mount Kenya. *Geogr. J.*, **58**: *9—25.*

AUTHUR, J. W., 1923. A Sixth Attempt on Mount Kenya. *Geogr. J.*, **62**: *205—209.*

AUTHUR, J. W., 1936. Radiation and anthocyanin pigments. In DUGGAR, B.M. Biological Effects of Radiation, **2**: 1109—1150.

ASDELL, S. A., 1946. Patterns of Mammalian Reproduction. New York.

ASTLEY MABERLEY, C. T., 1960. Animals in East Africa. Cape Town.

BERGHEN, VAN DEN, C., 1953. Quelques hépatiques récoltées par O. Hedberg sur les Montagnes de l'Afrique Orientale. *Svensk. bot. Tidskr.* **47**, 2. Uppsala.

BERGSTROM, E., 1955. British Ruwenzori Expedition, 1952. Glaciological Observations, Preliminary Report. *J. Glaciol.*, **2**, No. 17: *469—476.*

BOUGHEY, A. S., 1955. The nomenclature of the vegetation zones on the Mountains of Tropical Africa. *Webbia*, **XI**.

BRAESTRUP, F. W., 1941. A Study of the Arctic Fox in Greenland. *Medd. Grønland*, **131**: *1—101.*

BROOKS, C. E. P., 1949. Climate through the Ages. London.

BRUCE, E. A., 1934. The Giant Lobelias of East Africa. *Kew Bull. 61—88, 274.*

CAGNOLO, FR. C., 1933. The Akikuyu, Nyeri, Kenya.

CARCASSON, R. H., 1964. A preliminary survey of the zoogeography of African Butterflies. *E. A. Wildlife J.*, **2**: *122—157.*

CHAPER, R., 1886. Constatation de l'existence du terrain glaciaire dans l'Afrique equatoriale. *C. R. Acad. Sci.* C. **11**: *126—128.*

CHAPIN, J. P., 1923. Ecological aspects of bird distribution in Tropical Africa. *Amer. Nat.*, **57**: *106—125.*

CHAPIN, J. P., 1934. Up Kenya in the Rains. *Nat. Hist.*, **33**: *596—606;* **34**: *83—94.*

CHARNLEY, F. E., 1960. Some observations on the glaciers of Mount Kenya. *J. Glaciol.* **3**: *483—492.*

CHIRA, P., 1959. Personal communication.

CHOLNOKY, B. J., 1960. Diatomeen aus einem Teiche am Mt. Kenya in Mittelafrika. *Österr. bot. Z.*, **107**, 3/4: *351—365.*

COE, M. J., 1959. I.G.Y. Mount Kenya Expedition: Biological Report to Leader, Prof. I. S. Loupekine. (Unpublished).

COE, M. J., 1961. Notes on *Nectarinia johnstoni* on Mount Kenya. *The Ostrich*, Sept. 1961: *101—103.*

COE, M. J., 1962. Notes on the habits of the Mount Kenya Hyrax (*Procavia johnstoni mackinderi* Thomas). *Proc. Zool. Soc. Lond.*, **138**, 4: *639—644.*

COE, M. J., 1963. The Flora and Fauna of Mount Kenya and Kilimanjaro. East African Literature Bureau, Nairobi.

COE, M. J., 1966a. The biology and Breeding behaviour of *Tilapia grahami* Boulenger in Lake Magadi, Kenya. *Acta Tropica.*

COE, M. J., 1966b. Biogeography and the Equatorial Mountains. Palynology in Africa. 9th Report. (In Press).

COETZEE, J. A., 1964. Evidence for a considerable depression of the vegetation belts during the Upper Pleistocene on the East African mountains. *Nature, Lond.* **204,** No. 4958: *564—566.*
COOLEY, W. D., 1852. Inner Africa Laid Open. London.
COTTON, A. D., 1944. The Megaphytic habit in the tree Senecio and other genera. *Proc. Linn. Soc.,* Session **156** (2): *158—168.*
DALE, I. R. & GREENWAY, P. J., 1961. Kenya Trees and Shrubs. Nairobi.
DAUBENMIRE, R. F., 1959. Plants and Environment. John Wiley, N.Y.
D'ERMAN, M., 1949. Osservazioni su Recenti Fenomeni di Ritiro nei Ghiaccia del Monte Kenya. *Atti XIV Congr. geogr. ital. tenuto a Bologna:* 8.12. *354—356.*
DOWER, K. C. G., 1935. New Lake on Mount Kenya. *Geogr. J.,* **86:** *455—459.*
DRUMMOND, H., 1888. Tropical Africa. London.
DUTTON, E. A. T., 1929. Kenya Mountain. London.
ENGLER, A., 1904. Plants of the Northern Temperate Zone in their transition to the high mountains of Tropical Africa. *Ann. Bot.,* **18:** *523—540.*
FINCKH, L., 1902. Über die Gesteine des Kenya und des Kilimanjaro. *Cbl. Mineral. 204—205.*
FLINT, R. F., 1957. Glacial and pleistocene Geology. New York.
FLINT, R. F., 1959. Pleistocene climates in Eastern and Central Africa. *Bull. geol. Soc. Amer.,* **70,** 3: *343—374.*
FRIES, TH. C. E., 1923a. Die Alchemilla-Arten des Kenia, Mount Aberdare und Mount Elgon. *Ark. Bot.:* **18** (11). Uppsala.
FRIES, TH. C. E., 1923b. Beiträge zur Kenntnis der Flora des Kenia, Mt. Aberdare und Mt. Elgon (1) *Notizbl. Bot. Gart. Berl.,* **8:** *389—423.*
FRIES, R. E. & FRIES, TH. C. E., 1948. Phytogeographical researches on Mount Kenya and Mt. Aberdare, British East Africa. *K. Svenska Vetensk. Akad. Handl.* **111,** 25(5). Stockholm.
GEDGE, E., 1892. A recent exploration, under Captain F. G. Dundas, R. N. Up the Tana River to Mount Kenya. *Proc. Roy. geogr. Soc. N.S.,* **14:** *513—533.*
GIBB, J., 1950. The breeding biology of the Great and Blue Titmice. *Ibis,* **92:** *507—539.*
GIEGER, R., 1957. The Climate near the ground. Harvard.
GILLETT, J. B., 1955. The Relation between the Highland Floras of Ethiopia and British East Africa. *Webbia,* **XI.**
GODWIN, H., 1956. The History of the British Flora.
GREGORY, J. W., 1894a. Contributions to the Physical Geography of British East Africa. *Geogr. J.,* **4:** *289—315, 408—424, 505—524.*
GREGORY, J. W., 1894b. The glacial geology of Mount Kenya. *Quart. J. geol. Soc.,* **50:** *515—530.*
GREGORY, J. W., 1894c. An Expedition to Mount Kenya, *Fourtn. Rev., 327—337.*
GREGORY, J. W., 1896. The Great Rift Valley. London.
GREGORY, J. W., 1900. Geology of Mount Kenya. *Quart. J. geol. Soc.,* **56:** *205—222.*
GREGORY, J. W., 1921. The Rift Valleys and Geology of East Africa. London.
GREGORY, J. W., 1930. Geology of Mount Kenya. Appendix 3 in DUTTON (1930).
HAMMEN, TH. VAN DER & GONZALES, E., 1960. Upper Pleistocene and Holocene Climate and vegetation of the "Sabana de Bogota" (Colombo, S. America). *Leidse Geol. Meded.* **25:** *261—315.*
HAUMAN. L., 1934. Les Lobelia géants des Montagnes du Congo Belge. *Mem. Inst. Col. Belge* **11:** *1—52.*
HAUMAN, L., 1935. Les Senecio arborescents du Congo. *Rev. Zool. Bot. Afr.,* **18:** *1—76.*

HAUMAN, L., 1955. La Région Afroalpine en phytogéographie centro-africaine. *Webbia*. **XI**.
HEDBERG, O., 1951. Vegetation belts of the East African Mountains. *Svensk. bot. Tidskr.*, **45**: *140—202*. Uppsala.
HEDBERG, O., 1952. Cytological studies in East African mountain grasses. *Hereditas*, **38**: *256—266*.
HEDBERG, O., 1954. Taxonomic studies on Afro-Alpine Caryophyllaceae. *Svensk. bot. Tidskr.* **48**: *199—210*.
HEDBERG, O., 1955a. Altitudinal zonation of the vegetation on the East African Mountains. *Proc. Linn. Soc. Lond.*, Session **165**, 1952–3. Pt. 2. June 1955.
HEDBERG, O., 1955b. A pollen-analytical reconnaisance in tropical East Africa. *Oikos* **5** (1954): *137—166*. Copenhagen.
HEDBERG, O., 1955c. Some taxonomic problems concerning the Afro Alpine Flora.
HEDBERG, O., 1957. Afro-Alpine Vascular Plants. *Symb. bot. Upsal.* **XV**, 1: *1—411*.
HEDBERG, O., 1961a. Modern Taxonomic Methods and the Flora of Tropical Africa. *C.R. IVe Reunion Plenière de l'A.E.T.F.A.T. 265—278*. Lisbon.
HEDBERG, O., 1961b. The Phytogeographical position of the Afro-alpine Flora. Recent Advances in Botany: *914—919*. Toronto.
HEDBERG, O., 1963. Ekologisk Specialisering i Den Afroalpine Floran. Statens Naturvetenskapliga Forskningsrad: *158—170*. Stockholm.
HEDBERG, O., 1964. Features of Afro-Alpine plant ecology. *Acta Phytogeogr. Svecica*. Uppsala. 144 p.
HEINZELIN DE BRAUCOURT, J., 1953. Les Stades du Récession du Glacier Stanley occidental (Ruwenzori, Congo Belge). *Explor. Parc. Nat. Albert* 2e serie. Fasc. **1**.
HEINZELIN DE BRAUCOURT, J. & MOLLARET M., 1936. Biotopes de haute altude Ruwenzori. 1 *Explor. Parc. Nat. Albert* 2e serie. Fasc. **3**: *1—31*. 6 Pl.
HOBLEY, C. W., 1892. Peoples, Places and Prospects in East Africa. *Geogr. J.*, **2**: *97—123*.
HÖHNEL, L. VON, 1894. Discovery of Lakes Rudolph and Stefanie, 1887—1888. 2 vols. London.
HOLLISTER, N., 1919. East African Mammals in the United States National Museum. *U.S. Nat. Mus. Bull.* **99**, Part 2.
HOLLISTER, N., 1924. East African Mammals in the United States National Museum. *U.S. Nat. Mus. Bull.* **99**, Part 3.
HOOKER, J. D., 1874. On the sub-Alpine vegetation of Kilimanjaro, East Africa. *J. Linn. Soc. (Botany)*, **14**: *141—146*.
HUTCHINS, D. E., 1909. Report on the Forests of British East Africa. London.
HUTCHINSON, G. E., 1957. A treatise on Limnology. Vol. I. New York.
HUXLEY, SIR J., 1961. The Conservation of Wild Life and Natural habitats in Central and British East Africa. U.N.E.S.C.O.
JACKSON, F. J. & SLATER, V. L., 1938. The Birds of Kenya Colony and the Uganda Protectorate. 3 Vols. London.
JEANNEL, R., 1950. Hautes montagnes d'Afrique. *Publ. Mus. Nat. Hist. Nat.* Suppl. No. 1. Paris.
JENNINGS, D., 1963. Geology of Mount Kenya. Guide Book to Mount Kenya and Kilimanjaro. (Ed. I. REID) pp. *19—31*. Nairobi.
JEX-BLAKE, M., 1948. Some Wild Flowers of Kenya. Nairobi.
KENDEIGH, S. C., 1952. Parent care and its evolution in birds. *Ill. Biol. Monog.* **20**.
KENYATTA, J., 1938. Facing Mount Kenya. London.
KLUTE, F., 1920. Ergebnisse der Forschungen am Kilimandscharo 1912. Berlin.

KLUIJVER, H. N., **1933**. Bijdrage tot de Biologie en de ecologie van den Spreeuw (Sternus vulgaris vulgaris Lin) gedurende zijn voortplantingstijd. *Versl. Meded. Plantenziektenk. Wageningen.* **69**: *1—145.*

KRAPF, J. L., 1860. Travels, Researches and Missionary Labours in East Africa. London.

KRAPF, J. L., 1882. Mount Kenia. *Proc. R. geogr. Soc.* MS. **4**: *747—753.*

KUIPER., 1960. (Description of *Pisidium artifex* Kuiper). *Arch. Moll.* **89**: *67—76.*

LACK, D., 1951. Population ecology in birds. A review. *Proc. Xth Int. Orn. Congr. 409—448.*

LACK, D., 1959. The Natural Regulation of Animal Numbers. London.

LIVINGSTONE, D. A., 1962. Age of deglaciation of the Ruwenzori Range, Uganda. *Nature, Lond.* 194, No. 4831: *859—860.*

LLANO, G. A., 1962. The Terrestrial Life of the Antarctic. *Scient. Amer.* Sept. 1962: *213—230.*

LÖNNBERG, E., Mammals collected by the Swedish Zoological Expedition. *Kungl. Sv. Vet. Akad. Handl.* **48**. 5.

LÖNNBERG, E., 1929. The development and distribution of the African fauna in connection with and depending on climatic changes. *Ark. Zool.* **21A**: *1—31.*

MACKINDER, H. K., 1900. A Journey to the Summit of Mount Kenya, British East Africa. *Geogr. J.* **15**(5): *453—486.*

MACKINDER, H. K., 1930. Mount Kenya in 1899. *Geogr. J.* **76**: *529—534.*

MACKWORTH—PRAED, C. W. & GRANT, C. M. B., **1955**. Birds of Eastern and North Eastern Africa. 2 Vols. London.

MEINERTZHAGEN, R., 1937. Some Notes on the Birds of Kenya Colony, with Especial Reference to Mount Kenya. *Ibis* **14**,**1**: *731—760.*

MANI, M. S., 1962. Introduction to High Altitude Entomology. London.

MANI, M. S., 1967. Biogeography and Ecology of High Mountain Insects. *Series entomol.* **3**

MEYER, H., 1891. Across East African Glaciers. Berlin.

MEYER, H., 1900. Der Kilimandjaro. Berlin.

MOREAU, R. E., 1933. Pleistocene climatic changes and the distribution of life in East Africa. *J. Ecol.* **21**: *415—435.*

MOREAU, R. E., 1935. A synecological study of Usambara, Tanganyika Territory, with particular reference to birds. *J. Ecol.* **23**: *1—43.*

MOREAU, R. E., 1938. Climatic classification from the standpoint of East African Biology. *J. Ecol.* **26**: *467—496.*

MOREAU, R. E., 1944. Mount Kenya. A contribution to the Biology and Bibliography. *J. E.A. Nat. Hist. Soc.*, **18**: *61—92.*

MOREAU, R. E., 1944. Kilimanjaro and Mount Kenya. *Tanganika Notes and Records* **18**: *1—32.*

MOREAU, R. E., 1947. Relations between number in brood, feeding rate and nestling period in nine species of birds in Tang. Territory. *J. Anim. Ecol.* **16**: *205—209.*

MOREAU, R. E., 1952. Africa since the Mesozoic, with particular reference to certain biological problems. *Proc. zool. Soc., London,* **121**: *869—913.*

MOREAU, R. E., 1963. Vicissitudes of the African biomes in the late Pleistocene. *Proc. Zool. Soc., London.* **141**, 2: *395—421.*

MOREAU, R., 1963b. African Ecology and Human Evolution. Ed. Howell and Bourliere. pp. *28—42.*

NEW, C. 1873. Life, Wanderings and Labours in Eastern Africa, London.

NILSSON, E., 1932. Quarternary glaciations and pluvial lakes in British East Africa. Central trykeriet. Stockholm.

NILSSON, E., 1935. Traces of ancient changes of Climate in East Africa. *Geogr. Ann., H.* 1—2.

NILSSON, E., 1940. Ancient changes of Climate in British East Africa and Abessinia. *Geogr. Ann.*, H. 1—2: *1—79.*
NILSSON, E., 1952. Pleistocene climatic changes in East Africa. *Proc. Pan-African Congr. on Pre-history: 45—54.*
ORDE-BROWNE, G. ST J., 1916. Mount Kenya and its People (Chuka). *J. Afr. Soc.* **15**: *225—233.*
ORDE-BROWNE, G. ST J., 1918. The South East Face of Mount Kenya. *Geogr. J.* **51**: *389—392.*
PEARSALL, W. H., 1950. Mountains and Moorlands. London.
PEARSALL, W. H., 1956. Report on an Ecological Survey of the Serengeti National Park, Tanganyika. *Fauna Preservation Soc. 1—69.*
PEARSALL, W. H., 1961. Survival in Drought. *New Scientist,* **12,** No. 362: *489—491.*
RICHARDS, C. & PLACE, J., 1960. East African Explorers. London.
RICHARDS, C., 1961. Some Historic Journeys in East Africa. London.
RAVEN, J. & WALTERS, M., 1956. Mountain Flowers. London.
ROSIWAL, A., 1891. Über Gestein aus dem Gebiete Usambara und dem Stephanie See. *Denkschr. Akad. Wiss. Wien.*, **58**: *465—550.*
ROSS, W. MACGREGOR, 1911. The Snow Fields and Glaciers of Kenya. *Pall Mall Magazine: 197—208, 463—475.*
ROSS, W. MACGREGOR, 1911. Two finds on Mount Kenya. *J. E.A. Nat. Hist. Soc. 11,* No. 3: *60.*
ROSS, R., 1955a. Some Aspects of the Vegetation of the sub-Alpine Zone on Ruwenzori. *Proc. Linn. Soc., London;* **165**: *136—140.*
ROSS, R., 1955b. Some Aspects of the Vegetation of Ruwenzori. *Webbia,* **XI**: *451—457.*
SALE, J. B., 1965. Aspects of the behaviour and Ecology of Rock Hyraces, (Genera Procavia and Heterohyrax). Ph. D. Thesis presented to University of London.
SALT, G., 1951. The Shira Plateau of Kilimanjaro. *Geogr. J.*, **117,** 2: *150—164.*
SALT, G., 1954. A Contribution to the Ecology of Upper Kilimanjaro. *J. Ecol.*, **42,** 2: *375—423.*
SCAETTA, H., 1933. Bioclimats; climats des associations et microclimats du haute montagne en Afrique Centrale Equatoriale. *J. d'Agron. Col.* Brussel.
SCHELPE, E. A. C. L. E., 1949. Mountain vegetation in Southern Africa. *Proc. Linn. Soc., London.* **161**: *50.*
SCHMAUS, A., 1938. Der Einfluß der Mäusjahre auf das Brutgeschäft unserer Raubvögel und Eulen. *Beitr. Z. Fortpfl. Biol. Vög.* **14**: *181—184.*
SCHÜZ, E., 1942. Bestandsregelnde Einflüsse in der Umwelt des Weißen Storchs (C. Ciconia). *Zool. Jb.* Abt. Syst. ök. Geogr. **75**: *103—120.*
SCOTT, R. M., 1962 (in RUSSELL, E. W. The natural resources of E. Africa. pp. *67—76.* Nairobi).
SHANTZ, H. L. & MARBUT, C. F., 1923. The vegetation and soils of Africa. *Res. Ser. Amer. geogr. Soc.* No. 13.
SIMON, N., 1962. Between the Sunlight and the Thunder. London.
SPINK, P. C., 1949a. Further Notes on the Kibo crater and glaciers of Kilimanjaro and Mount Kenya. *Geogr. J.* **LXXIX**: *210—217.*
SPINK, P. C., 1949b. The Equatorial Glaciers of East Africa. *J. Glaciol.*, **I,** 5: *277—282.*
STEENIS, C. G. G. J. VAN, 1934–1935. On the origin of the Malaysian Flora. *Bull. Jard. Bot. Buitenz.*, (3), **13**: *135—162, 289—417.*
STEENIS, C. G. G. J. VAN, 1962. The Mountain Flora of the Malaysian Tropics. *Endeavour,* **XXI,** 83—84: *183—193.*
STEVENSON-HAMILTON, J., 1937. South African Eden. London.
STIGAND, CAPT. C. H., 1909. The Game of British East Africa. London.
STONEMAN, C. T., 1932. Wanderings in Wildest Africa. London. *89—129.*

STUHLMANN, F., 1894. Mit Emin Pasha ins Herz von Africa. Berlin.
SWAN, L. W., 1961. The Ecology of the High Himalayas. *Scient. Amer.* October 1961: *68—78.*
SYNGE, P. M., 1937. The mountains of the moon. 221 p., London.
THOMAS, O., 1900. 4. List of Animals obtained by Mr. H. J. Mackinder during his recent expedition in Mount Kenya, British East Africa. *Proc. Zool. Soc. Lond. 173—180.*
THOMSON, J., 1885. Through Massai Land. London.
TORNQUIST, A., 1893. Fragmente einer Oxford Fauna von Mtaru. in Deutsch Ostafrica nach dem von Dr. Stuhlmann gesammelten Material. *Jahrb. Hamburg Wiss. Anst.*, **X**: *263—287.* 3 Pls.
TROLL, C. & WIEN, K., 1949. Der Lewisgletscher am Mount Kenya. *Geogr. Ann.* 1—4: *257—274.*
TROLL, C., 1958. Tropical Mountain Vegetation. *Proc. 9th Pacific Sci. Congr.* **20.**
TROLL, C., 1958b. Zur Physiognomik der Tropengewächse; Bonn 1958: *1—75.* 68 Pls.
TROLL, C., 1958c. Structure Soils, Solifluction and Frost Climates of the earth. *U.S. Army Snow, Ice and Perma Frost Res. Est. Trans.* **43.**
TROLL, C., 1959. Die tropischen Gebirge. *Bonner geogr. Abh.* **23**: *1—93.* 29 Pls.
TROLL, C., 1960. The relationship between climates and plant geography of the Southern cold temperate zone and of the tropical high mountains. *Proc. Roy. Soc. B.* **152**: *529—532.*
TRUFFAUT, R., 1957. From Kenya to Kilimanjaro. London.
VARDE, R. POTTIER DE LA, 1955. Mousses récoltées par M. le Dr. Olov Hedberg, en Afrique orientale, au cours de la mission Suédoise de 1948. *Ark. f. Bot.* Serie 2, **3**, **8**.
VAN DER HORST, C. J., 1941. On the size of the litter and the gestation period of *Procavia capensis. Science.* N.Y. **33**: *430—431.*
VALEE, POUSSIN, J. DE LA, 1933. Les Glaciers du Ruwenzori. *Ann. Soc. Sci. Belg.* **LIII**: *45—57.*
VEEN, R. VAN DER & MEIJER, G., 1959. Light and Plant Growth. Philips Tech. Lib. Holland.
VAN ZINDEREN BAKKER, E. M., 1962. A late glacial and post-glacial climatic correlation between East Africa and Europe. *Nature, Lond.* **194;** 4824: *201—203.*
VAN ZINDEREN BAKKER, E. M., 1962. Palynology in Africa. 7th Report, S.A. Coun. Sci. and Ind. Res.
VAN ZINDEREN BAKKER, E. M., 1964. A pollen diagram from Equatorial Africa, Cherangani, Kenya. *Geologie en Mijnbouw,* **43**, **3**: *123—128.*
VAN ZINDEREN BAKKER, E. M., 1965. Upper Pleistocene stratigraphy and ecology on the basis of vegetation changes in sub-saharan Africa. Wenner-Gren Foundation for Anthropological Research.
WALTERS, M. & RAVEN, J., 1956. Mountain Flowers. p. 44. London.
WASSINK E. C. & STOLWIJK, J. A. J., 1956. Effects of light quality on plant growth. *Ann. Rev. Plant. Phys.*, **7**: *373—400.*
WILLIAMS, J. G., 1951. *Nectarinia johnstoni:* a revision of the species, together with data on plumages, moults and habits. *Ibis.* **93**: *579—595.*
WILLIAMS, J. G., 1963. A Field Guide to the Birds of East and Central Africa. London.
WITTOW, J. B., 1959. The Glaciers of Mount Baker Ruwenzori. *Geogr. J.* **CXXV**, 3—4: *370—379.*
ZEUNER, F. E., 1945. The Pleistocene period, its climate, chronology and faunal successions. Roy. Soc. Lond.
ZEUNER, F. E., 1948. Climate and Early Man in Kenya. *Man.* **48.**
ZEUNER, F. E., 1949. Frost Soils on Mount Kenya and the relation of frost soils to Aeolian deposits. *J. Soil. Sci.*, **I**: *20—30.*

INDEX